FUNCTION
CIRCUITS

THE  ELECTRONICS SERIES
BURR-BROWN

Wong and Ott • **FUNCTION CIRCUITS**

Graeme • **DESIGNING WITH OPERATIONAL AMPLIFIERS**

Graeme • **APPLICATIONS OF OPERATIONAL AMPLIFIERS**

Tobey, Graeme, and Huelsman • **OPERATIONAL AMPLIFIERS**

FUNCTION CIRCUITS
Design and Applications

YU JEN WONG, M.S.E.E.
Senior Engineer, Function Circuits
Burr-Brown Research Corporation

WILLIAM E. OTT, M.S.E.E.
Senior Engineer, Monolithic Circuits
Burr-Brown Research Corporation

McGRAW-HILL BOOK COMPANY
New York   St. Louis   San Francisco   Auckland   Bogotá   Düsseldorf
Johannesburg   London   Madrid   Mexico   Montreal
New Delhi   Panama   Paris   São Paulo
Singapore   Sydney   Tokyo   Toronto

Library of Congress Cataloging in Publication Data

Wong, Yu Jen, 1945-
  Function circuits.

  (The BB electronics series)
  Includes bibliographical references and index.
  1. Electronic circuits.   I. Ott, William E.,
1945-   joint author. II. Title. III. Series.
TK7867.W66   621.3815'3   76-43060
ISBN 0-07-071570-X

Copyright © 1976 by Burr-Brown Research Corporation. All rights reserved.
Printed in the United States of America. No part of this
publication may be reproduced, stored in a retrieval system,
or transmitted, in any form or by any means, electronic,
mechanical, photocopying, recording, or otherwise, without
the prior written permission of the publisher.

1234567890 KPKP 785432109876

The information conveyed in this book has been carefully reviewed and
is believed to be accurate and reliable; however, no responsibility is assumed for the operability of any circuit diagram or inaccuracies in calculations or statements. Further, nothing herein conveys to the purchaser a
license under the patent rights of any individual or organization relating
to the subject matter described herein.

*The editors for this book were Tyler G. Hicks and Lester Strong,
the designer was Naomi Auerbach, and the production supervisor
was Teresa F. Leaden. It was set in Caledonia
by The Kingsport Press.*

Printed and bound by The Kingsport Press.

CONTENTS

Preface xi

1. FUNDAMENTALS OF OPERATIONAL AMPLIFIERS 1
   1.1 Basic Circuits with Ideal Operational Amplifiers 2
       1.1.1 Inverting amplifiers 2
       1.1.2 Noninverting amplifiers 5
   1.2 DC Limitations of Practical Amplifiers 7
       1.2.1 Input offsets 7
       1.2.2 Open-loop gain 12
       1.2.3 Output impedance 16
       1.2.4 Common-mode rejection 18
       1.2.5 Input impedance 21
   1.3 Dynamic Characteristics of Practical Amplifiers 23
       1.3.1 Frequency response and stability 23
       1.3.2 Slew-rate limit and full-power bandwidth 29
       1.3.3 Settling time 31
       1.3.4 Capacitive loads 33

vi Contents

  1.4 Additional Amplifier Circuits 34
   1.4.1 Summers 34
   1.4.2 Integrators and differentiators 37
   1.4.3 Instrumentation amplifiers 42

2. LOGARITHMIC CONVERSION 46

  2.1 Bipolar Transistor as a Fundamental Logarithmic Element 46
   2.1.1 Compensation of the bulk resistance error 48
   2.1.2 Transistor connection versus diode connection 50
  2.2 Logarithmic Amplifiers 52
   2.2.1 Phase compensation of log amplifiers 52
   2.2.2 Protection circuitry for the logging transistor 55
   2.2.3 Temperature effects in log amplifiers 57
   2.2.4 Requirements on the operational amplifier 59
   2.2.5 Antilog and log ratio amplifiers 61
   2.2.6 Current inverters 63
  2.3 Log Amplifier Specifications and Testing 65
   2.3.1 Accuracy and dynamic range 67
   2.3.2 Output voltage drift with temperature 67
   2.3.3 Calibrating the log amplifier 68
   2.3.4 Testing the log amplifier's dynamic performance 69

3. MULTIPLIERS, DIVIDERS AND MULTIFUNCTION CONVERTERS 71

  3.1 General Description and Multiplier Specifications 72
   3.1.1 Simple departures from the ideal multiplier 72
   3.1.2 Multiplier dc error specifications 73
   3.1.3 Dynamic performance 76
   3.1.4 A typical multiplier specification sheet 82
  3.2 Test Multipliers Using Crossplot Technique 82
  3.3 Multiplication Techniques 84
   3.3.1 General fundamentals and a comparison chart 85
   3.3.2 Quarter-square multiplier 87
   3.3.3 Pulse width/pulse height multiplier 90
   3.3.4 Triangle-averaging multiplier 93
   3.3.5 Variable-transconductance multiplier 96
   3.3.6 Log-antilog multiplier 102
  3.4 Converting Any One-Quadrant Multiplier to Four-Quadrant Operation 106
  3.5 Analog Dividers 107
   3.5.1 Multiplier-inverted divider 109
   3.5.2 Log-antilog divider 110
   3.5.3 Offsetting a one-quadrant divider to accept bipolar numerator voltages 113
  3.6 Multifunction Converters 113
   3.6.1 Theory of operation 114
   3.6.2 Practical multifunction converter circuit 116
   3.6.3 Multiplication and division 117
   3.6.4 Exponential functions 119
  3.7 Square Rooters 121
   3.7.1 Multiplier-inverted square rooter 122

3.7.2 Divider-converted square rooter 123
3.7.3 Use the multifunction converter as a precision square rooter 124

4. RMS-TO-DC CONVERSION ... 126

4.1 General AC-to-DC Conversion 126
    4.1.1 Peak-responding ac-to-dc converters 127
    4.1.2 Average-responding ac-to-dc converters 128
    4.1.3 Why use the rms value? 129
4.2 Computing RMS Converters 131
    4.2.1 Implicit-computing rms converter 132
    4.2.2 Low-frequency errors 137
    4.2.3 Step response and settling time 139
    4.2.4 Exact rms conversion over a time interval t_o 141
    4.2.5 High-frequency limitations 143
4.3 Thermal RMS Converters 144
    4.3.1 Monolithic thermoelement 146
    4.3.2 Midband and dc transfer function 149
    4.3.3 Low-frequency errors 153
    4.3.4 Step response and settling time 156
    4.3.5 High-frequency limitations 160
    4.3.6 Input protection for thermal rms converters 160
4.4 Applying RMS Converters 162
    4.4.1 Reducing the output ripple 162
    4.4.2 Input ranging 164
    4.4.3 High-crest-factor inputs 169

5. POWER SERIES AND FUNCTION GENERATORS ... 177

5.1 Power Series and Its Analog Realization 177
5.2 Tchebysheff Polynomials 179
    5.2.1 Expansion into a series of Tchebysheff polynomials 182
    5.2.2 Computer-aided expansion into Tchebysheff polynomials 186
5.3 Arbitrary Function Generators 190
    5.3.1 Computer-aided n-degree polynomial generation 192
    5.3.2 Thermocouple linearization example 196
    5.3.3 Inverse function generators 200
5.4 Approximation Using Noninteger Exponent 201
    5.4.1 Sine-function generator 201
    5.4.2 Cosine-function generator 203
    5.4.3 Arctangent and vector magnitude 204
5.5 Implicit Feedback and Four-Quadrant Sine Generator 207
5.6 Diode Function Generators 209
    5.6.1 Temperature-compensated diode function generator 209
    5.6.2 Theory of operation 211
    5.6.3 Function setup 211

6. SIMPLIFIED DESIGN OF ACTIVE FILTERS ... 213

6.1 Synthesis of Active RC Networks 214
    6.1.1 Second-order low-pass filters 214

viii   Contents

  6.1.2 Second-order high-pass filters 216
  6.1.3 Active bandpass filters 217
  6.1.4 Active band-reject filters 220
  6.1.5 The universal active filter 222
  6.1.6 Universal linear-voltage-tunable active filter 225
  6.1.7 All-pass filters 226
  6.1.8 Third-order active filters with one operational amplifier 228
  6.1.9 Computer-aided design of fourth-order low-pass filter with one operational amplifier 232
 6.2 Butterworth, Bessel, and Techebysheff Characteristics 237
 6.3 Estimating Filter Complexity 243
 6.4 Transformation of Transfer Functions 244
  6.4.1 From low-pass to high-pass 245
  6.4.2 From low-pass to bandpass 246
  6.4.3 From low-pass to band-reject 247
 6.5 A Practical Multistage Active Filter Design Example 247
  6.5.1 Given specifications 248
  6.5.2 What if it is not geometric-mean symmetrical? 248
  6.5.3 Allowing guard bands 249
  6.5.4 Estimating number of required poles 249
  6.5.5 Normalizing the frequencies 249
  6.5.6 Finding the low-pass prototype 249
  6.5.7 Transformation from low-pass to bandpass 250
  6.5.8 Synthesis using the UAF 250
  6.5.9 Tuning the filter 251
 6.6 Computer-aided Analysis of Active Filters 252
  6.6.1 An example: eight-pole low-pass Butterworth design with two operational amplifiers 255

7. SPECIAL-PURPOSE APPLICATIONS OF FUNCTION CIRCUITS 260

 7.1 Voltage-controlled Function Circuits 260
  7.1.1 Voltage-controlled waveform generator 260
  7.1.2 Voltage-controlled quadrature oscillator 261
  7.1.3 Voltage-variable time constant 264
  7.1.4 Voltage-controlled exponentiator and antilog amplifier 264
 7.2 Modulators and Frequency Doublers 267
  7.2.1 Balanced modulator 267
  7.2.2 Amplitude modulator 268
  7.2.3 Frequency doubler 269
 7.3 Phase-Angle Detector 270
 7.4 Phase-Locked Loop 271
 7.5 Polar-to-Rectangular Resolver 272
 7.6 Automatic Gain Control Circuits 273
  7.6.1 Simple AGC using two-quadrant dividers 274
  7.6.2 Wideband automatic gain control 275
 7.7 Gain Measurement 275
  7.7.1 Wideband gain measurement 276
  7.7.2 Low-frequency gain error and nonlinearity measurement 278
 7.8 Power Measurement 280
  7.8.1 Single-phase power measurement 280

7.8.2 Three-phase power measurement   281
  7.8.3 Wideband power measurement   283
7.9 Mass Gas Flow Computation   283
Index   287

PREFACE

In the last decade a class of analog circuits has emerged which act as basic building blocks with which to perform a variety of instrumentation, computation, and control functions. These building blocks, including amplifiers, logarithmic converters, multipliers, dividers, rms converters, and active filters, are grouped under the generic term *function circuits*.

The availability of these relatively complex functions as precise, versatile building blocks has greatly simplified analog-circuit design, while also broadening the scope of practical analog circuits. Many of the principles of design and applications of these circuits were discussed in the McGraw-Hill/Burr-Brown book *Operational Amplifiers: Design and Applications*.[1] More recently, further design and applications information was put forward in another McGraw-Hill/Burr-Brown book, *Applications of Operational Amplifiers: Third-Generation Techniques*.[2]

However, the focus of these books was on the application of operational amplifiers. The design and application of the more complex function circuits, if they were treated at all, were treated as applications of operational amplifiers and were naturally somewhat abbreviated. This book was written to provide a complete treatment of the design and application,

---

[1] G. E. Tobey, J. G. Graeme, and L. P. Huelsman, *Operational Amplifiers: Design and Applications*, McGraw-Hill Book Company, New York, 1971.
[2] J. G. Graeme, *Applications of Operational Amplifiers: Third-Generation Techniques*, McGraw-Hill Book Company, New York, 1973.

as well as theory and testing, of the total array of modern analog-function circuits. An effort has been made to produce a practical reference which is of sufficient scope and breadth to be useful in preparing efficient circuit designs for each individual requirement. The common circuit configurations, their design equations, error sources, and relative merits are all discussed.

The first chapter of this book is a review of the fundamentals of operational amplifiers. It is meant to serve as a handy reference on the more common operational amplifier characteristics and applications for both the experienced analog-circuit designer and those new to the field. However, those new to analog design may benefit from a study of the two abovementioned reference books, and if extensive analog design is to be performed, they will be found valuable by both the experienced and those new to the field.

Chapters 2, 3, and 4 focus on the primary function circuits as components. The theory of operation, design, specifications, and testing of logarithmic amplifiers, multipliers, dividers, and rms converters are thoroughly treated. In considering these function circuits, emphasis is placed on such practical considerations of their design and applications as the effects of temperature, frequency stabilization, bandwidth optimization, and input and output protection to ensure reliable high accuracies in adverse environments.

The major applications of these function circuit components are treated in the following three chapters. Arbitrary function generators are covered in Chapter 5; computer programs are presented which ease the task of obtaining the most efficient analog realizations for function generators and signal linearization. Chapter 6 treats active filters, with an emphasis on those design techniques which are of practical values to the working engineer. Computer programs are also presented to relieve the designer of the laborious task of lengthy computations. Voltage-controlled circuits, such as automatic gain control, modulators, and demodulators, are covered in Chapter 7. Circuits for performing measurements and converting signals to dc levels are also presented in this last chapter. These include real power measurement for both single-phase and three-phase, amplifier gain accuracy, and linearity measurement.

The authors wish to acknowledge the extensive theoretical and practical background derived from the Burr-Brown environment. We are particularly grateful to F. Leland Payne for this thorough review of the manuscript and the many resulting improvements. Special thanks are also due to Fran Baker for her accurate and expeditious typing of the manuscript.

Yu Jen Wong
William E. Ott

1
FUNDAMENTALS OF OPERATIONAL AMPLIFIERS

Operational amplifiers are extremely versatile devices whose applications span the broad electronics industry. They have come to be used as a component, a basic building block of analog circuits. Operational amplifier designs have been so optimized that virtually any performance characteristic is readily available in a standard product from a number of manufacturers. Some of these optimizations provide low-cost, general-purpose performance, others provide ultra-high precision with very low drift and very high gain, still others provide high-power capabilities with high output voltage and current, and the list can go on to include the more complex isolation amplifiers which provide total input isolation from the output. The manufacturers of operational amplifiers provide extensive applications information, and numerous books are available on the design and application of operational amplifiers.[1-3] The purpose of this chapter is not to provide a comprehensive treatment of the design and application of operational amplifiers, but merely to provide the basic fundamentals essential to understanding the more complex function circuits discussed in the following chapters. This chapter can also be used as a handy reference for some more common application formulas and graphs.

1.1 Basic Circuits with Ideal Operational Amplifiers

The operational amplifier is simply a high-gain, direct-coupled amplifier, usually designed to amplify signals over a wide frequency range. They are normally used with external feedback networks to produce precise transfer characteristics which depend almost entirely on the feedback network. Most have a differential input, but some have only a single input terminal. Almost all have only a single output terminal. Operational amplifiers are generally represented by the symbol shown in Fig. 1.1. This symbol does not include the power-supply connections, which are assumed, unless they have a particular significance in the circuit. Also, single-ended amplifiers may be treated as a special case where the noninverting, plus input is grounded.

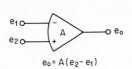

Fig. 1.1 Symbol for an operational amplifier. $e_o = A(e_2 - e_1)$

A preliminary analysis of feedback circuits using operational amplifiers is often simplified by assuming the amplifier has certain "ideal properties." While these properties are never realized in practice, the accuracy of the circuit analysis based on them is often adequate, and will almost always provide the insight needed to do a more rigorous analysis including the nonideal characteristics important in a particular application. The properties of the idealized amplifier which are usually assumed are

Gain = ∞	$A \to \infty$
Offset voltage = 0	$V_{OS} \to 0 (e_o = 0$ when $e_1 = e_2)$
Input bias current = 0	$I_{B1} = I_{B2} \to 0$
Input impedance = ∞	$Z_i \to \infty$
Output impedance = 0	$Z_o \to 0$
Bandwidth = ∞	$f_T \to \infty$
Phase shift = 0	$\vartheta \to 0$

The basic principles of analyzing feedback circuits will be demonstrated in this section with the analysis of the most common operational amplifier feedback circuits: inverting and noninverting amplifiers.

1.1.1 Inverting amplifiers The basic inverting amplifier is shown in Fig. 1.2. In this example, resistors are used for the input and feedback elements. R_1 is commonly referred to as the *input resistor* or *summing resistor*, and R_2 is referred to as the *feedback resistor*. If the operational amplifier is assumed to have ideal properties, no current flows into the input of the amplifier, and therefore $i_1 = i_2$:

$$i_1 = \frac{e_i - e_d}{R_1} = i_2 = \frac{e_d - e_o}{R_2}$$

Rearranging this yields the output

$$e_o = -\frac{R_2}{R_1} e_i + \left(\frac{R_2}{R_1} + 1\right) e_d \qquad (1\text{-}1)$$

But by definition

$$e_o = -A e_d \quad \text{or} \quad e_d = -\frac{e_o}{A}$$

and for arbitrarily large gain ($A \to \infty$) e_d is zero. The output then is simply

$$e_o = -\frac{R_2}{R_1} e_i \qquad (1\text{-}2)$$

Note that with an ideal operational amplifier the transfer function is only dependent on the feedback network. Also, because the amplifier has arbitrarily large gain, the summing point voltage e_d approaches zero and is a virtual ground.

Two summing point restraints result from the ideal properties:

1. No current flows into either input of the ideal operational amplifier.
2. When negative feedback is applied around the ideal operational amplifier, the differential input voltage approaches zero.

These two principles will be used repeatedly throughout this book, usually without being stated, when analyzing feedback circuits. Practical operational amplifiers do not meet the ideal properties. However, their deviations from the ideal are frequently insignificant, and the effect on the output can generally be treated in a separate analysis, often by relating the deviation from the ideal to the input as an equivalent input error.

The summing point restraints allow an immediate solution of the inverting summer shown in Fig. 1.3. Since there can be no voltage between the inputs of the operational amplifier and no current flows in its inputs, the input currents i_1, i_2, and i_3 can be written directly.

$$i_1 = \frac{e_1}{Z_1} \qquad i_2 = \frac{e_2}{Z_2} \qquad i_3 = \frac{e_3}{Z_3}$$

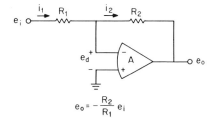

Fig. 1.2 Basic inverting amplifier circuit.

4   Function Circuits

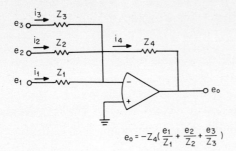

Fig. 1.3   Inverting summer.

The sum of these currents flows in the feedback element Z_4, generating the output voltage:

$$e_o = -Z_4\left(\frac{e_1}{Z_1} + \frac{e_2}{Z_2} + \frac{e_3}{Z_3}\right) \tag{1-3}$$

Thus the circuit is an inverting, summing amplifier, where each input can be operated on by separate scale factors.

The feedback networks do not have to be single components, such as resistors or capacitors, but may be complex elements composed of linear and nonlinear elements. The following chapters frequently apply nonlinear feedback elements to perform a variety of functions, and a slightly more complex linear feedback element is used in Fig. 1.4. Here a resistive-T feedback element is used, which has the advantage that a very

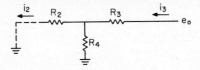

Short-circuit transfer impedance = Z_{sc}

$$Z_{sc} = \frac{e_o}{i_2} = \frac{R_2 R_3 + R_2 R_4 + R_3 R_4}{R_4}$$

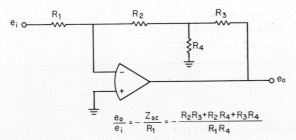

$$\frac{e_o}{e_i} = -\frac{Z_{sc}}{R_1} = -\frac{R_2 R_3 + R_2 R_4 + R_3 R_4}{R_1 R_4}$$

Fig. 1.4   Very high effective feedback impedances can be achieved with a T network of relatively low-value resistors.

high effective feedback resistance can be obtained without using such high-value resistors.

1.1.2 Noninverting amplifiers The inverting amplifiers discussed in the previous section may be realized with either single-ended or differential input amplifiers. The noninverting amplifier, on the other hand, requires the operational amplifier to have a noninverting input for signals and an inverting input for feedback. Usually such operational amplifiers are dif-

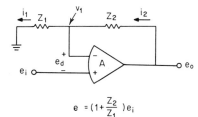

Fig. 1.5 Basic noninverting amplifier circuit.

ferential input types; however, there are amplifiers which operate only in the noninverting mode. The general noninverting circuit is shown in Fig. 1.5. The input signal is applied to the noninverting, plus input, and a portion of the output is fed back to the inverting input.

If it is assumed that the operational amplifier is ideal, $i_1 = i_2$, and then the following equations apply:

$$v_1 = \frac{Z_1}{Z_1 + Z_2} e_o \quad \text{for } I_B \to 0$$

$$v_1 = e_i + e_d$$

$$= e_i - \frac{e_o}{A} = e_i \quad \text{for } A \to \infty$$

Combining these two equations yields the output voltage:

$$e_o = \left(1 + \frac{Z_2}{Z_1}\right) e_i \tag{1-4}$$

Note that with an ideal operational amplifier, the noninverting circuit's transfer characteristics are only a function of the feedback network as it was in the inverting mode. A special case of the noninverting circuit is to make $Z_2 = 0$ or $Z_1 = \infty$; the gain is then exactly 1. In practice this voltage follower is implemented by making $Z_1 = \infty$ and simply shorting the output to the inverting input ($Z_2 = 0$); however, as will be seen in following sections, Z_2 may be made equal to the signal source impedance to cancel the effects of the input bias current of actual operational amplifiers.

A further modification of the noninverting amplifier is the difference

6 Function Circuits

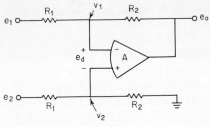

Fig. 1.6 Simple difference amplifier.

$$e_o = \frac{R_2}{R_1}(e_2 - e_1)$$

amplifier shown in Fig. 1.6. The transfer function is easily determined by using the law of superposition and the summing point restraints for ideal operational amplifiers. The law of superposition says that the total output is the sum of the outputs due to each input with the other input equal to zero:

$$e_o = e_o\bigg|_{e_2=0} + e_o\bigg|_{e_1=0}$$

With $e_2 = 0$ the circuit simplifies to the simple inverting amplifier previously discussed, and the output is found by inspection.

$$e_o\bigg|_{e_2=0} = -\frac{R_2}{R_1} e_1$$

When $e_1 = 0$ the circuit simplifies to the noninverting amplifier, except that the input has been attenuated to v_2 before being applied to the noninverting input of the operational amplifier. Because of the summing point restraint that $e_d = 0$, $v_1 = v_2$, and the output is

$$e_o\bigg|_{e_1=0} = \left(1 + \frac{R_2}{R_1}\right)\frac{R_2}{R_1 + R_2} e_2$$
$$= \frac{R_2}{R_1} e_2$$

Combining the output due to each input e_1 and e_2, the overall output is obtained:

$$e_o = \frac{R_2}{R_1}(e_2 - e_1)$$

Note that with an ideal operational amplifier, the output is determined entirely by the ratio of the feedback resistance to the summing resistance times the difference between the two input voltages; hence the name *difference amplifier* is used.

1.2 DC Limitations of Practical Amplifiers

Analysis of feedback circuits using operational amplifiers is generally simplified by assuming the amplifier has ideal properties, and even low-cost integrated-circuit operational amplifiers approximate the ideal operational amplifier to the extent that many applications can be adequately analyzed by assuming the amplifier has ideal properties. However, many applications will require levels of performance that may be limited by some of the amplifier's characteristics, and the ability to determine the significance and exact effect of the various operational amplifier characteristics is essential to determining which of the many amplifiers available will provide the required performance. The characteristics of operational amplifiers can be divided into two categories: dc characteristics and ac or dynamic characteristics. This section covers the dc characteristics. Common test configurations and typical values of the dc characteristics will be discussed, along with their effect on the output of the basic circuits.

1.2.1 Input offsets

Practical operational amplifiers have both an input offset voltage V_{OS} and input bias currents, I_{B1} in the inverting input and I_{B2} in the noninverting input. The output of the ideal operational amplifier is zero when its inputs are shorted. Primarily because of imperfect matching of the input devices, an actual amplifier's output voltage will not be zero when its inputs are both at zero potential. There is an equivalent dc voltage at the input of the amplifier, referred to as the *input offset voltage*. A further deviation from the ideal properties is the input bias currents. While no current flows in the inputs of the ideal operational amplifier, there are dc bias currents in the inputs of practical amplifiers. These bias currents are generally the base current of bipolar transistor input amplifiers or the gate leakage current of FET input amplifiers. The input offset voltage and bias currents are frequently the primary error sources; they can be accounted for by the model in Fig. 1.7.

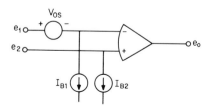

Fig. 1.7 Model of the input offset voltage and bias currents.

8 Function Circuits

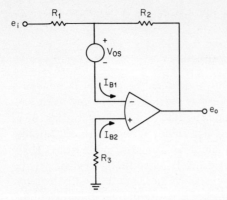

$$e_o = -\frac{R_2}{R_1} e_i \; (1+\frac{R_2}{R_1}) V_{OS} + R_2 I_{B1} - (1+\frac{R_2}{R_1}) R_3 I_{B2}$$

Fig. 1.8 Basic inverting amplifier including the input offsets. A bias current cancelling resistor R_3 has been included.

To determine the effect of these offsets on the output of the inverting amplifier, consider the circuit in Fig. 1.8. Perhaps the simplest method of determining the output of this circuit is to utilize the law of superposition; letting the offsets be zero,

$$e_o(e_i) = -\frac{R_2}{R_1} e_i$$

Letting the input be zero and considering the effect of each offset,

$$e_o(V_{OS}) = \left(1 + \frac{R_2}{R_1}\right) V_{OS}$$
$$e_o(I_{B1}) = R_2 I_{B1}$$
$$e_o(I_{B2}) = -\left(1 + \frac{R_2}{R_1}\right) R_3 I_{B2}$$

Now the total output can be written:

$$e_o = -\frac{R_2}{R_1} e_i + \left(1 + \frac{R_2}{R_1}\right) V_{OS} + R_2 I_{B1} - \left(1 + \frac{R_2}{R_1}\right) R_3 I_{B2} \quad (1\text{-}5)$$

If the amplifier is used noninverting by applying the input to R_3 rather than R_1 and grounding R_1 rather than R_3, the output is found to be

$$e_o = \left(1 + \frac{R_2}{R_1}\right) e_i + \left(1 + \frac{R_2}{R_1}\right) V_{OS} + R_2 I_{B1} - \left(1 + \frac{R_2}{R_1}\right) R_3 I_{B2} \quad (1\text{-}6)$$

Comparing Eq. (1-2) with Eq. (1-5) and comparing Eq. (1-4) with Eq. (1-6) yields the output of both circuits as

$$e_o = \text{(ideal output)} + \left(1 + \frac{R_2}{R_1}\right) V_{OS} + R_2 I_{B1} - \left(1 + \frac{R_2}{R_1}\right) R_3 I_{B2}$$

The input bias currents I_{B1} and I_{B2} are usually about the same, and their difference, the input difference current $I_{OS} = I_{B1} - I_{B2}$, is often specified. If R_3 in Fig. 1.8 is made equal to the parallel combination of R_1 and R_2, the input bias current error can be reduced to only the difference current,

$$e_o = \text{(ideal output)} + \left(1 + \frac{R_2}{R_1}\right) V_{OS} + R_2 I_{OS} \qquad (1\text{-}7)$$

The input offset voltage of general-purpose operational amplifiers is typically 1 mV for bipolar input types and 5 mV for FET input amplifiers. High-performance amplifiers have input offsets of 100 μV. These offsets are frequently the major source of error; a 1 percent of full scale error results from offset voltage for gains as low as 100 V/V with general-purpose amplifiers and a 10 V full scale. The input bias currents are also a major source of error. Bipolar input amplifiers have bias currents as high as 1 μA, and 50 nA is typical of many high-performance amplifiers. The input difference current is typically only one-tenth of the absolute bias current, or 100 nA for low-cost general-purpose amplifiers and 5 nA for high-performance amplifiers. FET input amplifiers generally achieve input bias currents of 10 pA, and 0.1 pA is available with ultra-low bias current amplifiers. The bias currents of the FET amplifiers are also usually matched; difference currents of only 20 percent of the absolute bias current are common. Input bias current errors are frequently significant where high impedances are necessary, particularly if high gain is also required.

Most operational amplifiers have provisions for offset voltage adjustment. However, some amplifiers do not make available the point for internal offset adjustment, and beyond that, there is frequently an interaction between the internal offset voltage compensation and the offset voltage drift versus temperature, the input bias current, and the input difference current. In order to avoid these interactions, offset voltage compensation external to the amplifier is useful.[2] Compensation of the input offsets with the inverting circuit is quite straightforward: simply place a compensating voltage between the noninverting input and common. While the effects of both the offset voltage and the bias current can be compensated by the offset voltage adjustment, because of the temperature effects it is best to match the impedance presented to both inputs of the amplifier as shown in Fig. 1.9. Typically, the difference current I_{OS} is only 10 percent of the absolute bias currents I_{B1} and I_{B2}, and the temperature sensitivity of I_{OS} is also 10 times better than that of I_{B1} and I_{B2}.

External offset voltage compensation for noninverting circuits is not as convenient. It is particularly difficult to compensate the offset voltage of unity-gain noninverting circuits. Offset null techniques for both the noninverting amplifier and the unity-gain follower are shown in Fig.

10 Function Circuits

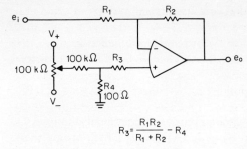

Fig. 1.9 The error in the inverting circuit due to the input offsets is easily compensated by applying a voltage to the noninverting input of the amplifier.

$$R_3 = \frac{R_1 R_2}{R_1 + R_2} - R_4$$

1.10.[2] In both cases the amplifier's offset voltage is compensated by summing a dc level with the input signal. This does modify the gain equation as shown, and requires the unity-gain follower to actually have a gain slightly greater than 1. Fortunately, the offset voltage effects are frequently insignificant when the amplifier is used with low gains, and when the gain is high the offset compensation does not produce significant gain errors.

By using either the internal offset adjustment or an external offsetting circuit such as those discussed, the offset voltage and bias current errors can be nulled. However, because of the noise associated with these offsets, and particularly because of their variation with temperature, power-supply voltage, and time, even under laboratory conditions these errors can only be reduced to within about 2°C of their temperature

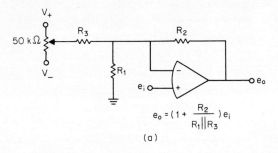

$$e_o = \left(1 + \frac{R_2}{R_1 \| R_3}\right) e_i$$

(a)

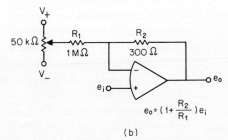

$$e_o = \left(1 + \frac{R_2}{R_1}\right) e_i$$

(b)

Fig. 1.10 External offset compensation of the noninverting circuit is achieved by summing a dc level with the input.

sensitivity. Typical general-purpose operational amplifiers have offset voltage drifts with temperature of 2 to 5 $\mu V/°C$, low-cost FET input amplifiers typically drift 5 to 20 $\mu V/°C$, but low-drift amplifiers are available with temperature sensitivities of 0.1 $\mu V/°C$ or less. The bias current of bipolar transistor input operational amplifiers typically drifts about 1.0%/°C, while FET input amplifiers' bias current typically doubles every 10°C. The input difference current typically has the same sensitivity, 1.0%/°C for bipolar units and double/10°C for FET units. The offset sensitivities to power-supply variations are typically 5 to 50 $\mu V/V$ and 1%/V for the offset voltage and bias currents, respectively. In addition to these drift sensitivities, the input offsets also have a noise component. The peak-to-peak and rms offset voltage noise is typically of the order of the offset voltage temperature sensitivity, i.e., 2 to 5 μV for general-purpose amplifiers. The input bias current shot noise is theoretically related to the base or gate current,[1]

$$i_n = \sqrt{2qI_B \, \Delta f}$$

where q is the charge of an electron and Δf is the noise bandwidth. However, because of a $1/f$ component of bias current noise, this level generally is not achieved.

The offset voltage, and its associated sensitivities, is easily measured by placing the amplifier in a high-gain configuration with zero input signal and measuring the amplifier's output. A typical circuit is shown in Fig. 1.11. A gain of 1,000 V/V is convenient, and the offset measured at the output is then 1 V/mV. The temperature sensitivity is easily measured by placing the amplifier in a constant-temperature chamber and measuring the output at the desired temperature extremes. The average drift versus temperature, $\Delta V_{OS}/\Delta T$, is defined as

$$\frac{\Delta V_{OS}}{\Delta T} = \frac{|V_{OS}(T_1) - V_{OS}(25°C)| + |V_{OS}(T_2) - V_{OS}(25°C)|}{|T_1 - T_2|}$$

where the specified temperature range is T_1 to T_2. The offset voltage sensitivity to other parameters, such as power-supply voltage, can be evaluated by using a similar expression.

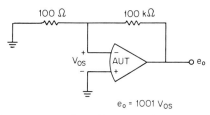

Fig. 1.11 The input offset voltage is easily measured by putting the amplifier in a high-gain configuration with zero inputs.

12   Function Circuits

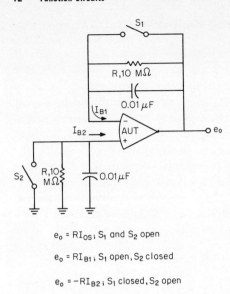

$e_o = RI_{OS}$; S_1 and S_2 open

$e_o = RI_{B1}$; S_1 open, S_2 closed

$e_o = -RI_{B2}$; S_1 closed, S_2 open

Fig. 1.12 The input bias currents are measured in the follower circuit, with high-value resistors used to develop an output proportional to the bias current and much greater than the offset voltage.

The input bias currents can be measured with the amplifier in a follower configuration and high-value resistors in each input. A suitable circuit is shown in Fig. 1.12. The resistors are shunted by capacitors to reduce noise, particularly 60 Hz pickup. This test configuration relies on the output due to the bias current flowing in the resistors being much greater than the offset voltage. With high-performance amplifiers, particularly FET input amplifiers, it may be necessary to first null the input offset with both switches closed to obtain good accuracy. The resistors can be made larger, providing a larger output due to bias current, to reduce the error due to the offset voltage. However, the test time will become longer due to the bypass capacitors. The capacitor values can be reduced to speed the test if careful shielding is used to prevent 60 Hz pickup.

1.2.2 Open-loop gain   The open-loop gain is the magnitude of the amplification factor of the operational amplifier. It is generally abbreviated as simply A. The closed-loop gain is commonly abbreviated A_{CL}. While the amplifier's gain at direct current and low frequencies is very large, typically 90 dB or 3×10^4 V/V for general-purpose amplifiers and 140 dB or 10^7 V/V for high-precision ones, it is finite and will produce an error in the closed-loop response. In order to analyze the effect of finite open-loop gain on the closed-loop response, it is convenient to define a feedback factor β:

$$\beta = \frac{1}{1 + Z_2/Z_1} \qquad (1\text{-}8)$$

In this case the output of the inverting amplifier, Eq. (1-1), can be written as

$$e_o = -\frac{Z_2}{Z_1}e_i + \frac{e_d}{\beta} \qquad (1\text{-}9)$$

where the more general summing and feedback impedances Z_1 and Z_2 have been substituted for the resistances R_1 and R_2. By definition $e_d = -e_o/A$, and by substituting this into Eq. (1-9) and rearranging, we obtain the closed-loop transfer function for the inverting circuit,

$$A_{CL} = \frac{e_o}{e_i} = -\frac{Z_2}{Z_1}\left(\frac{1}{1 + 1/A\beta}\right) \qquad \text{inverting circuit} \qquad (1\text{-}10)$$

A similar analysis of the noninverting circuit yields

$$A_{CL} = \frac{e_o}{e_i} = \left(1 + \frac{Z_2}{Z_1}\right)\left(\frac{1}{1 + 1/A\beta}\right) \qquad \text{noninverting circuit} \quad (1\text{-}11)$$

The term $A\beta$ is commonly referred to as the *loop gain* since it may be thought of as the gain around the loop formed by the amplifier and the feedback network. Comparing Eq. (1-2) with Eq. (1-10), and also comparing Eq. (1-4) with Eq. (1-11), shows the gain for both circuits to be

$$A_{CL} = \frac{e_o}{e_i} = \frac{\text{ideal gain}}{1 + 1/A\beta} \qquad (1\text{-}12)$$

In most operational amplifier circuits the loop gain is very large, and the expression for the closed-loop gain can be approximated as

$$A_{CL} = \frac{e_o}{e_i} = (\text{ideal gain})\left(1 - \frac{1}{A\beta}\right) \qquad (1\text{-}13)$$

The gain error is then simply $1/A\beta$. With even low-cost IC operational amplifiers, the open-loop gain is typically 3×10^4 V/V, and unless the feedback factor is small, the gain error will be negligible. For noninverting circuits β is the inverse of the ideal closed-loop gain, and if β is small, it is also approximately the inverse of the ideal closed-loop gain of the inverting circuit. Therefore, as would be expected, finite open-loop gain is generally a significant error source only for large closed-loop-gain applications. In that case, the gain error is approximately A_{CL}/A, since the feedback factor is essentially the inverse of the ideal closed-loop gain if the error is small.

As an example, suppose an inverting gain of 500 V/V is required; with a minimum open-loop gain of 3×10^4 V/V, the maximum error in the closed-loop gain will be 1.67 percent. While it is possible to compensate for the finite open-loop gain by a slight adjustment of the elements of the feedback network, the variation of the open-loop gain with temperature, supply voltage, load, and other external variables places a practical limitation on the ultimate gain accuracy achievable by this technique. It is usually best to choose an operational amplifier which has sufficient open-loop gain to provide the necessary gain accuracy without special adjustments.

For the more complex feedback networks the above relationships require some simple modifications. For the inverting summer shown in Fig. 1.3, the feedback factor is

$$\beta = \frac{1}{1 + Z_4/Z_{iT}}$$

where Z_{iT} is the parallel impedance of Z_1, Z_2, and Z_3. If the closed-loop gain is much greater than unity, the gain error is approximately $(A_{CL1} + A_{CL2} + A_{CL3})/A$, where A_{CL1}, A_{CL2}, and A_{CL3} are the ideal closed-loop gains for each input. For example, if the gain on e_1 is 150, on e_2 it is 30 and on e_3 it is 400; the gain error will be 1.93 percent if the amplifier's open-loop gain is 3×10^4 V/V. The output will appear as though the feedback resistor were 1.93 percent low.

The open-loop gain at direct current, A_0, is commonly specified; however, it is usually measured with an ac signal in order to discriminate the signal from the amplifier's dc offset. Provided that the measurement frequency is below the first pole frequency in the amplifier's open-loop gain (amplifier frequency response is discussed in Sec. 1.3), the measured gain will equal the dc gain. A straightforward gain measurement can be made by applying a small ac voltage to one of the amplifier's inputs and measuring the resulting ac output signal. In this case, the gain is simply the output divided by the input. The major problem with this technique is that the amplifier's input offset voltage is amplified by the open-loop gain, resulting in output saturation with most amplifiers. The offset voltage can be compensated by a voltage divider from the power supplies, and a suitable circuit is shown in Fig. 1.13a. While monitoring the amplifier's output with an oscilloscope, it is possible to adjust the potentiometer so that the output is continuously within the linear output range.

The offset voltage compensation can be avoided by supplying negative feedback at direct current but not at the test frequency. This can be accomplished by using a capacitor as the summing element, as shown in Fig. 1.13b. The feedback resistor R provides unity gain on the input offset voltage, preventing output saturation. The closed-loop response to the input, from Eq. (1–11), is

$$A_{CL} = \frac{e_0}{e_i} \doteq 1 + \frac{sRCA/(A+1)}{1 + sRC/(A+1)} \qquad \text{for } A \gg sRC$$

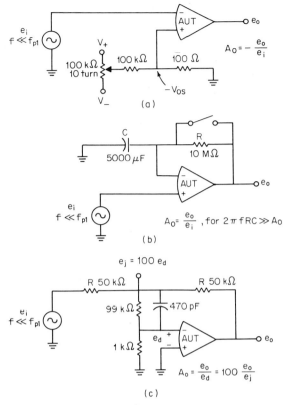

f_{o1} = amplifier's first pole frequency

Fig. 1.13 The low-frequency, or dc, open-loop gain can be measured in open-loop configurations (a) and (b), or in a closed-loop configuration such as (c) which provides a known output suitable for measuring the gain under particular load conditions.

Letting $s = j2\pi f$, and assuming $A_{\text{CL}} \gg 1$, the magnitude of the closed-loop gain is found to be

$$\left|A_{\text{CL}}\right| = \frac{2\pi f RCA/(A+1)}{\sqrt{1 + [2\pi fRC/(A+1)]^2}} \qquad (1\text{--}14)$$

From this it can be seen that for $2\pi fRC \gg A$, the closed-loop gain approaches the open-loop gain at that frequency:

$$\left|A_{\text{CL}}\right| = A \bigg|_{2\pi fRC \gg A}$$

Inspection of Eq. (1–14) shows that for a 1 percent measurement accuracy, $2\pi fRC$ only needs to be seven times the open-loop gain. While the feedback resistor does provide unity gain on the offset voltage, it is necessary for the summing capacitor C to charge to this dc voltage. Because of the large values of both the feedback resistor and the summing capacitor required to permit gain measurement at low frequencies, the time required to charge the capacitor to the offset voltage level may be long. A shorting switch in parallel with the feedback resistor can be closed to charge the capacitor very rapidly, then opened to make the gain measurement.

Both the open-loop-gain measuring circuits discussed above require that the input signal be very small, and so most signal generators will require an additional attenuator at their output; -40 dB is common. In addition, the amplitude of the input may require adjustment for each amplifier tested because of the wide variation in open-loop gain. Individual level adjustment is particularly important if it is necessary to test the amplifier's gain with a particular output signal level and load. The open-loop gain can be tested in a closed-loop condition with a specified output level and load by measuring the small differential input signal at the amplifier's input. A unity-gain inverting circuit would be suitable. The amplifier's output level is then identical to the input and can be loaded if the open-loop gain with a specified load is required. A new summing junction can be made by placing a voltage divider between the junction of the summing resistor, the feedback resistor, and the amplifier's inverting input, as shown in Fig. 1.13c. In this manner the amplifier's open-loop gain is reduced by a specific amount, providing a larger error voltage to be measured. For the circuit in Fig. 1.13c, the amplifier's gain has been reduced by 40 dB; therefore, the open-loop gain is 100 times the output voltage divided by the new summing junction voltage e_j. According to Eq. (1–13), as long as the effective open-loop gain is greater than 200 V/V, the output will be within 1 percent of the input voltage e_i.

1.2.3 Output impedance Practical operational amplifiers do not have zero output impedance. In fact, typical general-purpose amplifiers have 3 kΩ output impedance at direct current, while high-current amplifiers have output impedances of a few ohms. Figure 1.14 shows a noninverting amplifier, including the open-loop output impedance Z_o of the operational

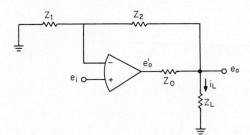

Fig. 1.14 Noninverting circuit, including the nonzero output impedance.

amplifier. This circuit can be analyzed to determine the effects of the open-loop output impedance. The resulting equations are

$$e'_o = A(e_i - \beta e_o) \qquad (1\text{-}15)$$

where

$$\beta = \frac{Z_1}{Z_1 + Z_2}$$

and

$$e_o = \frac{Z_L \parallel (Z_1 + Z_2)}{Z_o + Z_L \parallel (Z_1 + Z_2)} e'_o \qquad (1\text{-}16)$$

Alternatively, the output can be written as

$$e_o = A'(e_i - \beta e_o) \qquad (1\text{-}17)$$

where A' is the effective open-loop gain. Combining Eqs. (1-15), (1-16), and (1-17) yields

$$A' = \frac{Z_L \parallel (Z_1 + Z_2)}{Z_o + Z_L \parallel (Z_1 + Z_2)} A$$

$$= \frac{A}{1 + Z_o/Z_L + Z_o/(Z_1 + Z_2)} \qquad (1\text{-}18)$$

From this expression it is seen that the nonzero output impedance causes the effective open-loop gain to be reduced by the factor

$$\frac{1}{1 + Z_o/Z_L + Z_o/(Z_1 + Z_2)}$$

With a typical open-loop putput impedance of 3 kΩ at direct current and a total effective load of 2 kΩ ($Z_L \parallel (Z_1 + Z_2) = 2$ kΩ), the open-loop gain is reduced to 40 percent of its original value, or by 8 dB.

In addition to a reduced effective open-loop gain, the nonzero open-loop output impedance results in the closed-loop output impedance Z_{CL} also being greater than zero. By definition the closed-loop output impedance is

$$Z_{\text{CL}} = -\frac{\delta e_o}{\delta i_L} \qquad (1\text{-}19)$$

Also, the incremental load current flows through the output impedance, with the result that

$$\delta e'_o = \delta e_o + Z_o \, \delta i_L \qquad (1\text{-}20)$$

Differentiation of Eq. (1–15) yields

$$\delta e'_o = -A\beta\, \delta e_o \qquad (1\text{–}21)$$

Combining Eqs. (1–19), (1–20), and (1–21) and rearranging yields the closed-loop output impedance,

$$Z_{\text{CL}} = -\frac{\delta e_o}{\delta i_L} = \frac{Z_o}{A\beta + 1} \qquad (1\text{–}22)$$

For large loop gain, $A\beta \gg 1$, Eq. (1–22) shows the closed-loop output impedance to be the open-loop output impedance divided by the loop gain. For example, if the closed-loop gain is 500 V/V and the open-loop gain is 3×10^4 V/V, the open-loop output impedance will be reduced by a factor of about 60, resulting in a closed-loop output impedance of 50 Ω with a 3 kΩ open-loop output impedance.

The open-loop output resistance can be determined by measuring the change in open-loop gain due to a load. At direct current Eq. (1–18) simplifies to

$$A'_0 = \frac{R_L}{R_o + R_L} A_0$$

where R_o is the open-loop output resistance and R_L is the total effective load. For the open-loop-gain measuring circuits in Fig. 1.13b and c, the feedback resistors are a load on the amplifier in parallel with any additional load from the output to common. Generally, the feedback resistance in the open-loop-gain measurement circuit is much greater than the load and has a negligible effect. Then the output resistance is

$$R_o = \left(\frac{A_0}{A'_0} - 1\right) R_L$$

From this it is seen that it is convenient to adjust the load resistance for a 6 dB loss of open-loop gain ($A_0/A'_0 = 2$), in which case the load resistance is equal to the open-loop output resistance.

1.2.4 Common-mode rejection

Ideally the output of the operational amplifier responds only to the difference between the two inputs $(e_2 - e_1)$ and produces no output due to the common-mode voltage e_{icm}. The common-mode voltage is the average of the two input voltages:

$$e_{icm} = \frac{e_1 + e_2}{2}$$

where e_1 and e_2 are the inverting and noninverting input voltages, respectively. Practical amplifiers will respond to some extent to the common-mode voltage, producing an output voltage e_b, and a common-mode gain can then be defined:

$$A_{cm} = \frac{e_b}{e_{icm}}$$

The common-mode error can be referred to the input as the common-mode rejection ratio (CMRR), which is the ratio of the differential open-loop gain to the common-mode gain,

$$\text{CMRR} = \frac{A}{A_{cm}}$$

The CMRR has dimensions of volts per volt (V/V) but is frequently referred to in decibels,

$$\text{CMR (dB)} = 20 \log \text{CMRR (V/V)}$$

A model of an operational amplifier including the common-mode error is shown in Fig. 1.15.

With inverting circuits the common-mode voltage is zero and no error is produced because of the finite CMRR. However, with noninverting circuits a common-mode voltage does exist, and the CMRR may be the major error source. Figure 1.16 shows a noninverting amplifier, including the input common-mode error voltage. Note that since the open-loop gain is very high and the offset voltage is small, the common-mode input is essentially the voltage at the noninverting input terminal, e_i in this case. The common-mode error appears as simply an additional input, and the output can be written directly from Eq. (1–11):

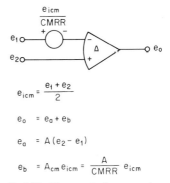

Fig. 1.15 The error in the output due to nonzero common-mode gain can be referred to the input as a common-mode rejection ratio.

$$e_o = \left(1 + \frac{R_2}{R_1}\right)\left(\frac{1 + 1/\text{CMRR}}{1 + 1/A\beta}\right) e_i$$

Thus, finite CMRR introduces an additional component of error in the closed-loop response. The magnitude of the common-mode rejection ratio of most operational amplifiers is typically about the same as, or somewhat less than, the amplifier's open-loop gain. Therefore, in low-gain configurations the common-mode error may be the major source of gain error and nonlinearity. This is particularly true of the voltage-follower configuration, where $R_2/R_1 = 0$ and $\beta = 1$.

20 Function Circuits

The common-mode rejection ratio can be tested in the difference amplifier circuit discussed in Sec. 1.1.2. The appropriate circuit is shown in Fig. 1.17. The common-mode voltage is

$$e_{icm} = \frac{R_{2b}}{R_{1b} + R_{2b}} e_i$$

and the amplifier's differential input voltage is

$$e_d = \frac{-e_o}{A} + \frac{e_{icm}}{\text{CMRR}}$$

The voltage at the inverting input of the amplifier is then

$$v_1 = \frac{R_{2b}}{R_{1b} + R_{2b}} \left(1 + \frac{1}{\text{CMRR}}\right) e_i - \frac{e_o}{A}$$

and the output voltage is

$$e_o = \frac{v_1 - e_i}{R_{1a}} (R_{1a} + R_{2a}) + e_i$$

Combining these expressions yields

$$e_o\left(1 + \frac{R_{1a} + R_{2a}}{R_{1a}A}\right) = \left[\frac{R_{2b}(R_{1a} + R_{2a})}{R_{1a}(R_{1b} + R_{2b})} - \frac{R_{2a}}{R_{1a}}\right] e_i + \frac{R_{2b}(R_{1a} + R_{2a})}{R_{1a}(R_{1b} + R_{2b})} \frac{e_i}{\text{CMRR}}$$

If the resistor ratios R_{1a}/R_{2a} and R_{1b}/R_{2b} are sufficiently matched, and the amplifier has a large open-loop gain compared to $(1 + R_2/R_1)$, the CMRR is found from this to be

$$\text{CMRR} = \frac{R_2}{R_1} \frac{e_i}{e_o}$$

The required resistor match can also be found from the above expression and is

$$\left|1 - \frac{R_{2a}}{R_{1a}} \frac{R_{1b}}{R_{2b}}\right| \leq \left|\frac{R_{1a} + R_{2a}}{R_{1a} \text{CMRR}}\right|$$

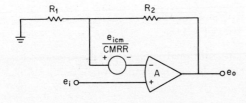

$e_{icm} = e_i$

$e_o = \left(1 + \dfrac{R_2}{R_1}\right) \left(\dfrac{1 + 1/\text{CMRR}}{1 + 1/A\beta}\right) e_i$

Fig. 1.16 Finite common-mode rejection produces an additional gain error in the noninverting circuit's transfer function.

Fundamentals of Operational Amplifiers

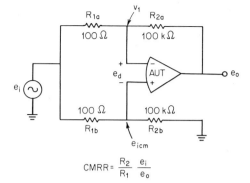

Fig. 1.17 An amplifier's common-mode rejection can be measured in the difference circuit with zero differential input.

$$\text{CMRR} = \frac{R_2}{R_1} \frac{e_i}{e_o}$$

The common-mode gain and the open-loop gain are frequently nonlinear functions of the common-mode voltage level, especially for FET input amplifiers. For this reason the CMRR must be tested over the full common-mode voltage range specified. The common-mode input range is defined as the peak value of common-mode input voltage that can be applied and still obtain the specified CMRR. Saturation of the input stage usually limits the common-mode voltage range, and this is readily detected in the CMRR test circuit of Fig. 1.17 by determining the input voltage level which results in a sudden, large increase in the output voltage.

1.2.5 Input impedance There are both a finite differential input impedance Z_{id} between each input of the operational amplifier and a com-

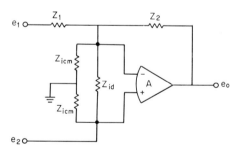

Fig. 1.18 There are both a finite differential input impedance between the inputs of practical operational amplifiers and a common-mode input impedance from each input to ground.

mon-mode input impedance Z_{icm} from each input to common. The effect of these finite input impedances on both inverting and noninverting circuits can be determined using the circuit in Fig. 1.18. Letting the input at the noninverting input of the amplifier be zero, $e_2 = 0$, the effect of the input impedance on the inverting circuit can be found. The effective input impedance of the amplifier is the parallel combination of the differential input impedance and the common-mode input impedance,

$$Z_i' = \frac{Z_{id} Z_{icm}}{Z_{id} + Z_{icm}}$$

The voltage drop on the input impedance Z_i' is simply the output divided by the open-loop gain, $-e_o/A$. An error current then flows in the input impedance which subtracts from the current in the feedback impedance Z_2. Thus an error in the output results. The total output is then the sum of the output with infinite input impedance and the error in the output due to the finite input impedance,

$$e_o \doteq e_o' \left(1 - \frac{Z_2}{AZ_i'}\right)$$

where e_o' is the output with infinite input impedance. Note that the finite input impedance results in a gain error and that the effective input impedance is multiplied by the amplifier's open-loop gain. This error is generally negligible; even for low-cost bipolar input amplifiers with differential input impedance as low as 300 kΩ and an open-loop gain of 3×10^4 V/V, the effective impedance is about 10^{10} Ω. FET input amplifiers with common-mode and differential input impedances of about 10^{10} Ω and open-loop gains of 3×10^4 V/V will have a totally negligible effective impedance AZ_i' of more than 1.5×10^{14} Ω.

The effect of finite input impedances on the performance of the noninverting circuit may be more significant and can be evaluated by letting e_1 be zero. It is immediately apparent that the common-mode input impedances limit the stage input impedance presented to the input signal e_2 to less than Z_{icm} and that the common-mode input impedance at the inverting input of the amplifier is in parallel with the closed-loop determining impedance Z_1. The voltage drop on the differential input impedance is again $-e_o/A$, and an error current then flows in it and the feedback impedance Z_2. The resulting output voltage is

$$e_o \doteq e_o' + \frac{Z_2}{Z_{icm}} e_2 - \frac{Z_2}{AZ_{id}} e_o$$

where e_o' is the output with infinite input impedance. Since $e_o = A_{CL} e_2$ and $e_o' \doteq e_o$, this can be rewritten as

$$e_o \doteq e_o' \left(1 + \frac{Z_2}{A_{CL} Z_{icm}} - \frac{Z_2}{AZ_{id}}\right)$$

As for the inverting circuit, the finite input impedances result in a gain error in the closed-loop response of the noninverting circuit which is usually negligible.

The more significant effect of the finite input impedances is the resulting input impedance presented to the input signal e_2. As previously stated, the common-mode input impedance is an obvious load on the input, but the differential input impedance is also a load on the input signal. The error current flowing in the differential input impedance is e_o/AZ_{id}, and this must be supplied by the input signal. The resulting overall closed-loop input impedance presented to the input signal e_2 is

$$Z_{iCL} = Z_{icm} \left\| \frac{A}{A_{CL}} Z_{id} \right.$$

For the noninverting circuit $1/A_{CL} = \beta$, and the input impedance can be rewritten as

$$Z_{iCL} = \frac{1}{1/Z_{icm} + 1/A\beta Z_{id}}$$

In most applications the finite input impedance of practical operational amplifiers has a negligible effect on the dc and low-frequency closed-loop gain. However, in addition to a resistive component of input impedance, practical amplifiers also have a capacitive component of input impedance. The input capacitance of most general-purpose amplifiers is about 3 pF, common-mode and differential, and this may be the dominant error source at higher frequencies.

1.3 Dynamic Characteristics of Practical Amplifiers

Many applications of operational amplifiers are strictly dc or low-frequency, and the error sources discussed in the previous section are all that are important. However, there are other applications of operational amplifiers where frequency response and the other dynamic characteristics are most important. This section will consider the most common ac or dynamic characteristics and limitations of practical operational amplifiers.

1.3.1 Frequency response and stability The effect of finite open-loop gain was discussed previously in Sec. 1.2.2, and for dc and low-frequency applications it is usually adequate to consider the open-loop gain to be constant and use the dc open-loop gain A_0 to determine the closed-loop response. However, for higher frequencies it is necessary to include the frequency dependence of the amplifier's open-loop gain. The frequency response of an operational amplifier is conveniently represented by a Bode plot. The absolute value of the voltage gain in decibels, $A(\text{dB}) = 20 \log_{10} A$ (V/V), is plotted versus frequency, using a decade logarithmic frequency scale. A typical operational amplifier Bode plot is shown in

Fig. 1.19. The intersections of the various segments of the linear approximations occur at the singularities in the amplifier's open-loop-gain response, and a mathematical approximation can be written directly:

$$A(s) = \frac{A_0 \, (1 + s\tau_{z1})}{(1 + s\tau_{p1})(1 + s\tau_{p2})(1 + s\tau_{p3})}$$

The closed-loop circuit cannot supply more gain than is available from the operational amplifier itself, so at high frequencies the closed-loop response intersects and follows the open-loop gain curve. The intersection point between the closed-loop and open-loop curves is important because the rate of closure between the two curves determines whether the closed-loop circuit will be stable. According to Bode's criterion, the rate of closure between the open-loop frequency response $A(j\omega)$ and the reciprocal of the feedback factor $1/\beta(j\omega)$ must be less than 12 dB per octave for a stable closed-loop system. To guarantee sufficient phase margin, most general-purpose operational amplifiers have internal phase compensation networks which produce approximately a constant -6 dB

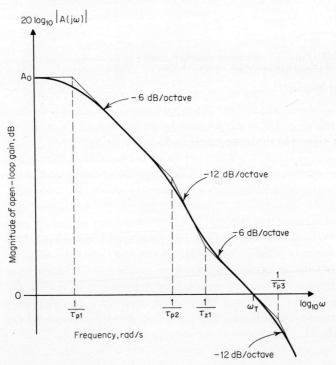

Fig. 1.19 Bode plot of a typical operational amplifier's open-loop-gain frequency response.

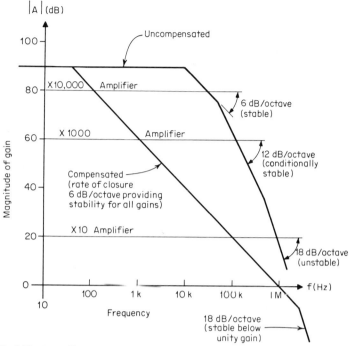

Fig. 1.20 Typically, operational amplifiers require phase compensation to obtain a single-pole response which ensures closed-loop stability for all feedback factors.

per octave rolloff in the amplifier's open-loop-gain response. Figure 1.20 shows the open-loop Bode plot of a typical general-purpose amplifier with compensation and also uncompensated. Several different closed-loop gains are shown to illustrate stable and unstable conditions.

Many operational amplifiers intended for wideband applications have external phase compensation. This allows the phase compensation to be tailored for each particular feedback condition, thereby providing the optimum frequency response for each application. Figure 1.21 illustrates this broadbanding principle where the closed-loop bandwidth has been improved from f_c to f'_c, while still maintaining stability. The broadbanded amplifier may be used for any gain level above the originally designed level, but it may not be used for lower gains without readjusting the compensation.

When the amplifier has a -6 dB per octave rolloff, its open-loop gain can be approximated as having one pole, and the mathematical expression for the gain simplifies to

$$A(s) = \frac{A_0}{1 + s\tau_p}$$

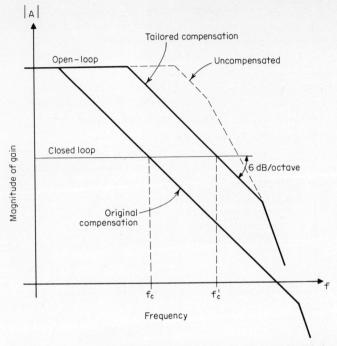

Fig. 1.21 By tailoring the amplifier's phase compensation for each application, the optimum gain-bandwidth product can be obtained.

Substituting this into Eq. (1-12), we obtain the closed-loop gain as a function of complex frequency s:

$$A_{\text{CL}}(s) = \frac{\text{ideal gain}}{1 + (1 + s\tau_p)/A_0\beta}$$

Rearranging, we obtain the closed-loop gain as a function of the closed-loop gain at direct current, A_{CL0}, and frequency;

$$A_{\text{CL}}(s) = \frac{A_{\text{CL0}}}{1 + s\tau_p/(A_0\beta + 1)}$$

For most applications the dc loop gain $A_0\beta$ is large, and the closed-loop gain can be written in terms of the dc closed-loop gain, the feedback factor, and the amplifier's gain-bandwidth product ω_T ($\omega_T = A_0/\tau_p$):

$$A_{\text{CL}}(s) = \frac{A_{\text{CL0}}}{1 + s/\beta\omega_T} \tag{1-23}$$

From this expression the closed-loop -3 dB bandwidth ω_c is found to be

$$\omega_c = \beta\omega_T$$

In Fig. 1.22 the simple inverting voltage amplifier is used to illustrate the significance of loop gain for both the gain magnitude and the phase shift of the closed-loop transfer function.

The plots in Fig. 1.22 illustrate several important points:

1. The corner frequency of the closed-loop circuit is the intersection of the open-loop gain curve $A(j\omega)$ and the reciprocal of the feedback factor $1/\beta$. For the noninverting amplifier, $1/\beta = A_{CL}$, and if the closed-loop gain is large, $1/\beta \doteq A_{CL}$ for the inverting amplifier also.
2. At low frequencies, where the loop gain is greater than 1, the closed-loop response is determined by the feedback network. At higher frequencies, where the feedback factor is greater than the open-loop gain, the closed-loop response approaches the open-loop gain.

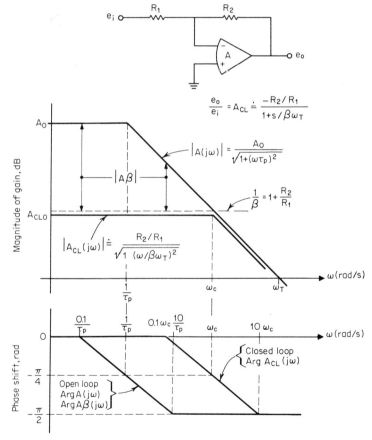

Fig. 1.22 Complete Bode plot of inverting amplifier, including both gain and phase-shift plots.

3. At low frequencies the loop gain is a constant and, as shown in Sec. 1.2.2, produces a gain error equal to the reciprocal of loop gain, $1/A\beta$. ($A\beta = 200$ produces approximately a 0.5 percent error in the closed-loop gain.)
4. At high frequencies the error due to finite loop gain is complex, producing a gain magnitude error and also phase shift which results in a vector error. The phase-shift error is approximately a linear function of loop gain, being approximately $1/A\beta(j\omega)$. The magnitude error, on the other hand, is smaller for frequencies below the closed-loop corner frequency. The magnitude error can be found directly from Eq. (1-23).

$$|A_{\text{CL}}(j\omega)| = \frac{|A_{\text{CL0}}|}{\sqrt{1 + (\omega/\beta\omega_T)^2}}$$

$$\text{Magnitude error} = 1 - \frac{1}{\sqrt{1 + (\omega/\beta\omega_T)^2}}$$

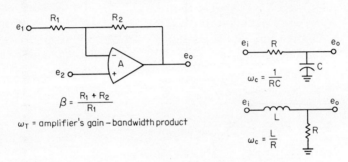

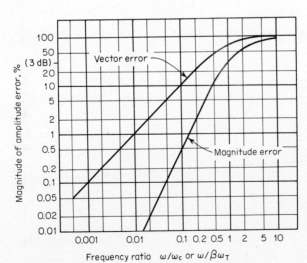

Fig. 1.23 The magnitude error of a single-pole stage is much less than the absolute, or vector, error. (The graph can be used for high-pass single-pole stages by inverting the frequency ratio to ω_c/ω.)

Both the vector error and the magnitude error of a single-pole transfer function are shown in Fig. 1.23. As is readily apparent, for 1 percent magnitude flatness the bandwidth can be less than one-fourteenth that required for a 1 percent absolute or vector accuracy.

1.3.2 Slew-rate limit and full-power bandwidth

In addition to the linear, or small-signal, frequency limitations discussed in the previous section, practical operational amplifiers also have a large-signal frequency limitation. The large-signal frequency limitation manifests itself in two forms: the slew-rate limit S_R, which is the maximum rate of change of the output voltage, and the full-power bandwidth f_p, which is the highest frequency at which the amplifier will deliver rated output without significant distortion. Usually both these limitations are due to the maximum rate at which the phase compensation capacitance can be charged. In many designs the total output swing appears across the phase compensation capacitor, which typically has a value of 30 pF in general-purpose amplifiers. Because of the design of high-gain differential amplifiers,[1,3] typically there is only about 30 μA of current available for charging this capacitor. Since the current through a capacitor is the capacitance times the rate of change of the voltage across the capacitor, $i = C\, dv/dt$, a slew-rate limit occurs.

$$I = C\,\frac{\Delta V}{\Delta t}$$

$$S_R = \frac{\Delta V}{\Delta t} = \frac{I}{C}$$

For the general-purpose amplifier, $I = 30$ μA and $C = 30$ pF, and the slew rate S_R is then 1 V/μs. Wideband amplifiers may have sufficient input stage currents for slew rates as high as 2000 V/μs to be obtained.

In general, the unity-gain voltage-follower configuration is the worst-case slew-rate condition, and slew rate is therefore measured in this configuration. Also, with some amplifiers the slew rate will be affected by the load on the amplifier, so that the slew rate is usually specified with rated load, over the rated output voltage range. Figure 1.24 shows an amplifier in this configuration along with the typical output waveform. The input is driven by a high-frequency square wave of sufficient amplitude to cause the output of the amplifier to swing beyond its specified range. The output is overdriven by an amount sufficient to cause the rounding due to finite small-signal bandwidth to occur outside the specified output range. The slew rate is the slope of the output transition between the specified limits. As illustrated in Fig. 1.24, the slew rate for positive transitions, S_{R+}, may not equal the slew rate for negative transitions, S_{R-}. The lower slew rate is commonly specified.

The slewing-rate limitation will not only cause distortion of the output for square-wave inputs, but also distort the output for sine-wave inputs of sufficiently high frequency. The slew rate limits the maximum slope of the sinusoidal output, and by equating the sine wave's maximum slope to the slew-rate limit, the full-power bandwidth is found:

$$e = E_p \sin 2\pi f t$$

$$\left|\frac{de}{dt}\right| = 2\pi f E_p \cos 2\pi f t$$

$$S_R = \left|\frac{de}{dt}\right|_{max} = 2\pi f_P E_p$$

$$f_P = \frac{S_R}{2\pi E_p} \tag{1-24}$$

where E_p is the peak amplitude of the sine wave. Generally, the full-power bandwidth is specified for rated output, which is typically ±10 V.

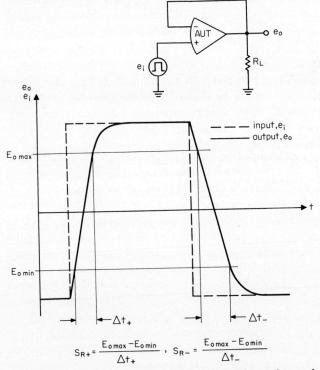

Fig. 1.24 The slew rate of practical operational amplifiers is frequently unsymmetrical, requiring testing for both positive and negative transitions.

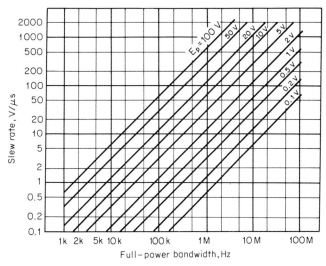

Fig. 1.25 Full-power bandwidth is theoretically related to slew rate, and is the primary frequency limitation when the output swing is large.

However, the full-power bandwidth for reduced output swing is frequently of most interest, and it can be found from Eq. (1–24), or from the graph in Fig. 1.25.

With the graphs in Figs. 1.23 and 1.25, the upper frequency limit of an operational amplifier circuit can be found. With the amplifier's slew-rate specification and the sine-wave amplitude required, the full-power bandwidth is determined from Fig. 1.25 or Eq. (1–24). Then, if the frequency of interest is less than the full-power bandwidth, the magnitude and vector errors can be found from Fig. 1.23. Note that if the frequency of interest is above the full-power bandwidth of the amplifier, the graph in Figure 1.23 does not apply, since it is generated for linear frequency response.

1.3.3 Settling time Settling time is the time required following application of an input step for the output to settle to within a specified error band of its final value. A complex output step response, containing overshoot and ring, is shown in Fig. 1.26. The settling time t_s is a strong function of the error-band η specified, particularly when there is output ring following a step input.

General-purpose amplifiers are usually phase-compensated to have a single-pole frequency response, as discussed previously in Sec. 1.3.1. Such an amplifier's output response to a step input $E_i u(t)$ can be found from the inverse Laplace transform of Eq. (1–23):

$$e_o(t) = A_{\mathrm{CL}} \left(1 - e^{-t\beta\omega_T}\right) E_i$$

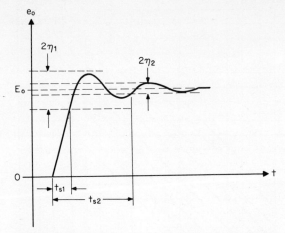

Fig. 1.26 If the step response of an amplifier rings, the settling time will be a particularly strong function of the settling error allowed.

The settling error η is

$$\eta = \frac{e_o(t_s) - e_o(\infty)}{e_o(\infty)} = e^{-t_s \beta \omega_T}$$

From this the settling time as a function of the settling error is found:

$$t_s = \frac{1}{\beta \omega_T} \ln \frac{1}{\eta} \qquad (1\text{-}25)$$

$\beta \omega_T$ is the corner frequency of the closed-loop response ω_c, allowing the settling time to be rewritten as

$$t_s = \frac{1}{\omega_c} \ln \frac{1}{\eta} \qquad (1\text{-}26)$$

The amplifier's slew rate may limit the settling time obtained. If the slewing time is of the order of, or particularly if it is greater than, the theoretical settling time, it will increase the actual settling time obtained. Further second-order effects, such as the complex condition of imperfect pole-zero cancellation in the open-loop gain frequency response, and internal thermal feedback because of changes in the output power, may also increase the actual settling time obtained. Generally, if the settling error is relatively large, such as 1 percent, the settling time will only be limited by the slew rate and the theoretical settling time given in Eq. (1-25). However, for small settling errors, of the order of 0.01 percent, the second-order effect will usually dominate, and the amplifier will require much more time to settle than Eq. (1–25) predicts.

1.3.4 Capacitive loads The addition of capacitive reactance at the output of an operational amplifier may lead to peaking and possibly instability at high frequencies. The capacitive load tends to break with the open-loop output impedance of the amplifier, introducing another pole in the open-loop-gain frequency response. As the rate of closure approaches 12 dB per octave, the closed-loop response begins to peak and possibly oscillate. This problem is most severe at low gain levels and at high frequencies.

In order to evaluate the significance of a particular capacitive load, the added pole's frequency should be determined. If it occurs at a lower frequency than the closed-loop corner frequency, instability may result. This would seem to be a straightforward calculation, $1/2\pi R_o C_L$. However, the output impedance is a function of frequency, which complicates the calculation. For general-purpose amplifiers, the high-frequency output impedance is of the order of 100 Ω, allowing the amplifier to be operated in unity gain with as much as a 1000 pF capacitive load without causing any oscillation. Wideband amplifiers will have as low as 25 Ω output resistance at high frequencies; however, their bandwidth is much greater, and they may only be stable with capacitive loads of less than 100 pF. If capacitive-load instability is a problem, the output of the amplifier can be decoupled from C_L with an external phase compensation network, as shown in Fig. 1.27. The phase compensation resistor R_3 should be made about equal to the high-frequency output impedance of the amplifier; 100 Ω is adequate with most amplifiers. C_1 should produce a response zero in conjunction with the feedback resistor R_2 at a frequency lower than the closed-loop corner frequency $\omega'_c = \beta \omega'_T$. ω'_T is less than ω_T because of the new pole at $1/2\pi(R_o + R_3)C_L$.

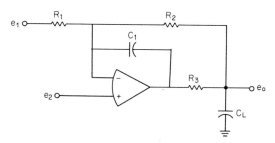

Fig. 1.27 Stability with capacitive loads can be maintained by decoupling the amplifier's output from the capacitive load at high frequencies.

1.4 Additional Amplifier Circuits

In addition to the basic inverting and noninverting amplifiers, there are several very common circuits which utilize operational amplifiers. Some of these which are frequently encountered in the following chapters will be discussed in this section. Specifically, the circuits to be discussed are general summers, integrators, and instrumentation amplifiers. Several other amplifier circuits for specific applications will be discussed in later chapters. In Chapter 2, Sec. 2.2.4, a simple FET buffer to be used in conjunction with a bipolar input operational amplifier is presented, and additional input offset compensating circuits are also discussed. A wideband input ranging amplifier useful in ac instrumentation is discussed in Chapter 4, Sec. 4.4.2. Also discussed in Sec. 4.4.2 are techniques for combining a high-precision low-drift operational amplifier with a wideband amplifier to obtain the best performance characteristics of each amplifier.

1.4.1 Summers[4] If the inverting and noninverting amplifiers previously discussed are combined, general summing amplifiers can be realized. Figure 1.28 shows a general summing amplifier which will take the sum or difference of any number of inputs. The equation of the output voltage is

$$e_o = a_{p1}e_{p1} + a_{p2}e_{p2} + \cdots - a_{n1}e_{n1} - a_{n2}e_{n2} - \cdots$$

where $e_{p1}, e_{p2}, \ldots$ are the inputs to be added, with scale factors $a_{p1}, a_{p2}, \ldots$, and $e_{n1}, e_{n2}, \ldots$ are the inputs to be subtracted, with scale factors $a_{n1}, a_{n2}, \ldots$.

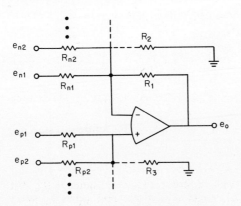

Fig. 1.28 A general summer can be built with a single operational amplifier.

$e_o = a_{p1}e_{p1} + a_{p2}e_{p2} + \cdots - a_{n1}e_{n1} - a_{n2}e_{n2} - \cdots$

Analysis of the circuit, assuming that the operational amplifier is ideal, yields the scale factors as

$$a_{pi} = \frac{R_{t3}}{R_{pi}} \left(\frac{R_{t2} + R_1}{R_{t2}} \right)$$

$$a_{ni} = \frac{R_1}{R_{ni}}$$

where the resistance R_{t2} is the parallel combination of R_2 and all the inverting input resistors $R_{n1}, R_{n2}, \ldots$:

$$R_{t2} = R_2 \parallel R_{n1} \parallel R_{n2} \parallel \cdots$$

and similarly, R_{t3} is the parallel combination of R_3 and all the noninverting input resistors $R_{p1}, R_{p2}, \ldots$:

$$R_{t3} = R_3 \parallel R_{p1} \parallel R_{p2} \cdots$$

Inspection of these equations shows that R_2 and R_3 are not always necessary; if the sum of 1 and the inverting scale factors is greater than the sum of the noninverting scale factors, R_3 must be included to attenuate the noninverting inputs; however, R_2 will be unnecessary. If the sum of the positive scale factors is greater than the sum of 1 and the negative scale factors, R_2 must be included to provide gain on the noninverting inputs, but R_3 is unnecessary. And finally, if the sum of the positive scale factors equals the sum of 1 and the negative scale factors, neither R_2 nor R_3 is necessary. To summarize;

$$R_2 = \infty \quad \text{for } \left(1 + \sum_{i=1}^{N_n} a_{ni}\right) > \sum_{i=1}^{N_p} a_{pi} \quad (1\text{-}27\text{a})$$

$$R_3 = \infty \quad \text{for } \sum_{i=1}^{N_p} a_{pi} > \left(1 + \sum_{i=1}^{N_n} a_{ni}\right) \quad (1\text{-}27\text{b})$$

$$R_2 = R_3 = \infty \quad \text{for } \left(1 + \sum_{i=1}^{N_n} a_{ni}\right) = \sum_{i=1}^{N_p} a_{pi} \quad (1\text{-}27\text{c})$$

where N_p and N_n are the number of positive and negative scale factors, respectively.

To determine the actual values of the resistors, the following step-by-step procedure should be used:

1. Decide what composite source resistance should be presented to the input terminals of the operational amplifier. This resistance is R_{t3}, and if the operational amplifier were ideal, its value would only be a function of the input impedance required and the degree of noise pickup

sensitivity tolerable. However, with a practical operational amplifier, the amplifier's input impedance is not infinite, nor is the amplifier's input bias current zero. As previously discussed in Sec. 1.2, the amplifier's input impedance is almost always sufficiently high not to produce a significant error; however, the input bias current does frequently introduce significant errors. This step-by-step procedure for determining the resistor values provides matched impedances to the inputs of the amplifier, and therefore minimizes the error due to input bias current. A value of 5 to 10 kΩ for R_{t3} provides low noise pickup, good bandwidth, and a reasonable input impedance for most applications.

2. Add up the negative scale factors and add 1 to determine $\left(1 + \sum_{i=1}^{N_n} a_{ni}\right)$.

3. Add up the positive scale factors to determine $\sum_{i=1}^{N_p} a_{pi}$.

4. Multiply the value of R_{t3} chosen in step 1 above by the larger of $\sum_{i=1}^{N_p} a_{pi}$ or $\left(1 + \sum_{i=1}^{N_n} a_{ni}\right)$; this is the required value of R_1. The multiple used is the reciprocal of the feedback factor ($1/\beta$) and is useful for determining the summer's gain accuracy and frequency response, as discussed in Secs. 1.2.2 and 1.3.1. It is also the gain on the equivalent input offset voltage, including the effects of the input difference current on R_{t3}.

5. The value of R_2 or R_3, if not determined to be infinite by Eq. (1–26), is

$$(R_2 \text{ or } R_3) = \frac{R_1}{\left|1 + \sum_{i=1}^{N_n} a_{ni} - \sum_{i=1}^{N_p} a_{pi}\right|}$$

6. The values of the summing resistors, $R_{n1}, R_{n2}, \ldots$ and $R_{p1}, R_{p2}, \ldots$, are all found by dividing R_1 by the associated coefficient:

$$R_{n1} = \frac{R_1}{a_{n1}} \quad R_{n2} = \frac{R_1}{a_{n2}} \cdots$$

$$R_{p1} = \frac{R_1}{a_{p1}} \quad R_{p2} = \frac{R_1}{a_{p2}} \cdots$$

As an example, suppose the required transfer function is

$$e_o = 15e_1 + 4.5e_2 - 7e_3$$

The sum of the negative scale factors plus 1 is 8, and the sum of the positive scale factors is 19.5. If the source impedance presented to the inputs of the operational amplifier is chosen as 5 kΩ, the feedback resistor R_1 should be 97.5 kΩ. Since the sum of the positive scale factors is greater

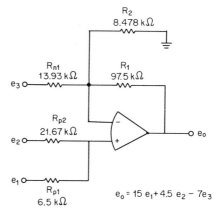

Fig. 1.29 This example summer, whose resistor values were determined with the step-by-step procedure, provides balanced source impedances to the amplifier's inputs, thereby reducing the effects of input bias currents.

than the sum of 1 and the negative scale factors, gain on the noninverting inputs will be required, and R_3 should be infinite. The required value of R_2 is found from step 5 above to be 8.478 kΩ. The summing resistors required are $R_{n1} = 13.93$ kΩ, $R_{p1} = 6.5$ kΩ and $R_{p2} = 21.67$ kΩ. The appropriate circuit is shown in Fig. 1.29.

1.4.2 Integrators and differentiators[1]   Integration and differentiation are frequently encountered operations in analog computations. Integrators, particularly, also have very wide application in signal conditioning, control, and instrumentation. Integrators will be frequently encountered in the following chapters, and an understanding of the practical analog integrator is essential. The basic integrator, the one most commonly used, is illustrated in Fig. 1.30. The current through the summing resistor R is

$$i_1 = \frac{e_i - e_d}{R}$$

Assuming the operational amplifier is ideal results in the capacitor current $i_2 = i_1$; therefore, the voltage drop on the feedback capacitor is

$$v = \frac{1}{RC} \int (e_i - e_d) \, dt$$

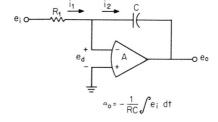

Fig. 1.30 A simple integrator requires only a summing resistor and a feedback capacitor; however, a summing integrator can be built by providing multiple summing resistors.

and the output voltage is then

$$e_o = e_d - \frac{1}{RC}\int (e_i - e_d)\,dt$$

By definition $e_d = -e_o/A$, and for arbitrarily large gain ($A \to \infty$) the amplifier's differential input voltage approaches zero ($e_d \to 0$). The resulting output voltage is then

$$e_o = -\frac{1}{RC}\int e_i\,dt$$

As previously discussed in Sec. 1.2.1, practical operational amplifiers have both an input offset voltage and input bias currents. These error sources can be accounted for in the integrator as shown in Fig. 1.31. The output of the integrator including these error sources is

$$e_o = \frac{-1}{RC}\int e_i\,dt + \frac{1}{RC}\int V_{os}\,dt + \frac{1}{C}\int I_B\,dt + V_{os}$$

The offset voltage produces both an input error, which is integrated along with the input signal, and a constant-output offset, which is simply the offset voltage. The amplifier's input bias current flows through the feedback capacitor, producing an error voltage on the capacitor which is proportional to the integral of the bias current. The error at the output due to the input bias current can often be significantly reduced by placing a resistor in series with the amplifier's noninverting input equal to the summing resistor R. In this manner, the error due to input bias current is reduced to only the input difference current, which is often only a tenth of the bias current. Since the offset voltage and bias currents are essentially constant, they produce a ramp error at the output of the integrator. With zero input the offsets will be integrated to produce a ramp output which will eventually cause the output to reach one or the other of its saturation levels.

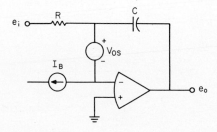

Fig. 1.31 Both the input bias current and the offset voltage produce a ramp error in the output of the integrator.

$$e_o = -\frac{1}{RC}\int e_i\,dt + \frac{1}{RC}\int V_{os}\,dt + \frac{1}{C}\int I_B\,dt + V_{os}$$

The frequency response of an ideal integrator is

$$\frac{E_o(s)}{E_i(s)} = \frac{1}{s\tau}$$

where τ is an arbitrary scale factor, $\tau = RC$ with the analog integrator. Because of the finite open-loop gain and bandwidth of practical operational amplifiers, the analog integrator's frequency response will deviate from the ideal at both extremes of the frequency spectrum. From Eq. (1–10), the transfer function of the inverting circuit, the integrator's frequency response is

$$\frac{E_o(s)}{E_i(s)} = -\frac{1}{sRC}\left(\frac{1}{1 + 1/A\beta(s)}\right)$$

If it is assumed that the amplifier has a single-pole response with a gain-bandwidth product of ω_T, the resulting frequency response of the integrator is

$$\frac{E_o(s)}{E_i(s)} = \frac{-A_0}{1 + s[(A_0 + 1)RC + A_0/\omega_T] + s^2 A_0 RC/\omega_T}$$

If it is further assumed that $A_0 \gg 1$ and $RC \gg 1/\omega_T$, the integrator's transfer function simplifies to

$$\frac{E_o(s)}{E_i(s)} = \frac{-A_0}{(1 + sA_0 RC)(1 + s/\omega_T)}$$

The Bode plots of both the ideal and the practical integrator are shown in Fig. 1.32, along with the amplifier's open-loop frequency response. Note that at low frequencies the practical integrator's closed-loop gain is naturally limited to the open-loop gain of the amplifier. Also, at the operational amplifier's unity-gain frequency the practical integrator's frequency response has an additional pole.

The inverse of integration is differentiation, and as might be expected, the basic circuit implementation of a differentiator is the inverse of that of the integrator. Figure 1.33 shows the basic circuit configuration of a differentiator. The controlling equations are

$$i_1 = C\frac{d(e_i - e_d)}{dt}$$

$$e_o = e_d - Ri_2$$

If the amplifier is assumed to have the ideal properties $i_1 = i_2$ and $e_d \to 0$, the output then becomes

$$e_o = -RC\frac{de_i}{dt}$$

40 Function Circuits

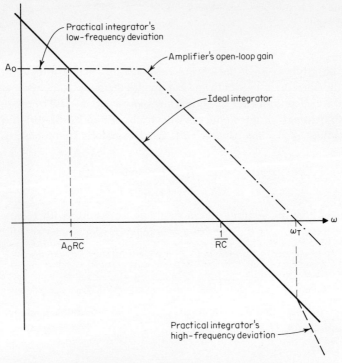

Fig. 1.32 A practical integrator has a limited gain at low frequencies, equal to the amplifier's open-loop gain, and an additional pole at high frequencies, occurring at the unity-gain frequency of the operational amplifier.

The frequency response of the ideal differentiator is

$$\frac{E_o(s)}{E_i(s)} = s\tau$$

where τ is an arbitrary scale factor, $\tau = RC$ with the analog differentiator. The Bode plot of a typical operational amplifier's frequency response along with the differentiator's frequency response is shown in Fig. 1.34.

At direct current the capacitor summing element has infinite impedance;

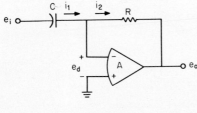

Fig. 1.33 The basic differentiator circuit configuration is simply the inverse of the integrator.

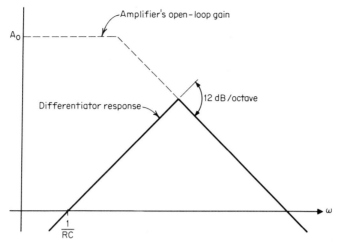

Fig. 1.34 The rate of closure where the differentiator frequency response curve intersects the open-loop-gain frequency response curve of practical operational amplifiers is 12 dB per octave, resulting in an inherently unstable circuit.

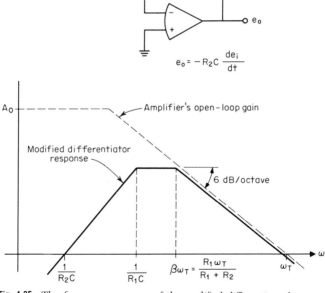

Fig. 1.35 The frequency response of the modified differentiator becomes simply that of an inverting amplifier at high frequencies, thereby avoiding stability problems.

hence the closed-loop gain is zero. At higher frequencies the impedance of the capacitor decreases, and the closed-loop gain increases at the rate of 6 dB per octave. Since the gain increases with frequency, differentiators are very susceptible to random noise. More importantly, as illustrated in Fig. 1.34, the rate of closure is about 12 dB per octave, making the simple differentiator inherently unstable. A practical method of reducing the noise and preventing instability is to include a stop resistor in series with the summing capacitor. This is illustrated in Fig. 1.35, and the modified differentiator's frequency response is also shown. At high frequencies the impedance of the capacitor is less than that of the stop resistor, and the high-frequency closed-loop response becomes that of the simple inverting amplifier with resistive elements.

1.4.3 Instrumentation amplifiers[1,2,5]   Instrumentation amplifiers, often referred to as differential amplifiers or data amplifiers, are closed-loop-gain blocks with differential inputs and an accurately predictable input-to-output relationship. They have very high input impedance and common-mode rejection. This makes them ideal for accurately amplifying low-level signals in the presence of large common-mode voltages, such as those from strain gage bridges, thermocouples, and other transducers. They can also be used to effectively eliminate ground loops.

The instrumentation amplifier differs fundamentally from the operational amplifier. It is designed to be used as a closed-loop gain block. The necessary feedback networks are usually contained within the amplifier, requiring only one external gain-setting resistor. The gain, input and output impedances, frequency response, and other characteristics are specified for the closed-loop, committed operation. Operational amplifiers, on the other hand, are open-loop devices whose closed-loop performance depends upon the external networks. Some instrumentation amplifiers have digitally programmable gain, in which case even the gain-setting resistors are contained within the amplifier. To operate IC instrumentation amplifiers, the user may have to supply one or two "feedback" resistors in addition to the gain-setting resistor. These user-supplied resistors do not affect the input impedance or the common-mode rejection as they would in operational amplifier circuits.

Ideally, the instrumentation amplifier responds only to the difference between two input signals and exhibits an extremely high input impedance, both differentially and common-mode. The output voltage is generally developed single-ended with respect to ground and is equal to the product of the amplifier gain and the differential input voltage:

$$e_o = A_d(e_2 - e_1)$$

Actual amplifiers will depart from these ideal characteristics and can be modeled as in Fig. 1.36. As illustrated in Fig. 1.36, the output voltage

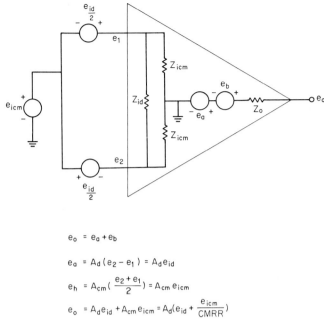

$$e_o = e_a + e_b$$
$$e_a = A_d(e_2 - e_1) = A_d e_{id}$$
$$e_b = A_{cm}\left(\frac{e_2 + e_1}{2}\right) = A_{cm} e_{icm}$$
$$e_o = A_d e_{id} + A_{cm} e_{icm} = A_d\left(e_{id} + \frac{e_{icm}}{CMRR}\right)$$

Fig. 1.36 Model of an instrumentation amplifier.

has two components: one component proportional to the differential input voltage e_{id}, and another proportional to the common-mode input voltage e_{icm}. The constant A_d is the differential gain (usually fixed by the external gain-setting resistor), and A_{cm} represents the common-mode gain of the amplifier. This is more commonly specified in terms of the common-mode rejection ratio, which is the ratio of the differential gain to the common-mode gain. The impedance Z_{id} is the differential input impedance, and the common-mode input impedance is represented by two equal components, Z_{icm}, from each input to ground. These finite input impedances will contribute an effective gain error due to loading of the source impedance, and will degrade the overall common-mode rejection of the amplifier stage if the source resistances are unbalanced. The nonzero output impedance Z_o will also produce a gain error whose value will depend on the load resistance.

A single operational amplifier can be used to build a differential, or difference, amplifier. This circuit was discussed previously in Sec. 1.1 and is illustrated in Fig. 1.6; it is suitable in some applications. However, if the application requires high gain, high input impedance, high CMR, or easily selectable gain over a wide range, this simple approach will probably be insufficient. The differential input impedance of this simple differential amplifier is that of the input resistors, which is generally low, particularly if high gain is required. Also, even though the operational

44   Function Circuits

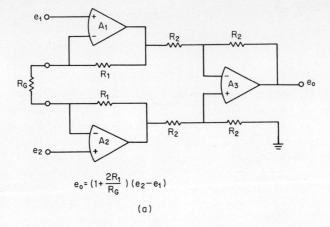

$$e_o = (1 + \frac{2R_1}{R_G})(e_2 - e_1)$$

(a)

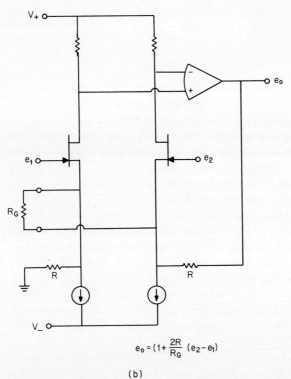

$$e_o = (1 + \frac{2R}{R_G})(e_2 - e_1)$$

(b)

Fig. 1.37 Instrumentation amplifiers can be built using a number of operational amplifiers as in (a), or using a differential stage and a single operational amplifier as in (b).

amplifier may have excellent CMR, the finite resistor matching will limit the overall CMR obtained. Possibly more significantly, if gain switching is required, two matching and tracking resistors must be switched in tandem.

A common configuration of modular instrumentation amplifiers consists of three operational amplifiers, as shown in Fig. 1.37a. The two input amplifiers provide a differential gain of $(1 + 2R_1/R_G)$ and a common-mode gain of unity. The output amplifier A_3 is a unity-gain differential amplifier. Resistors R_1 do not significantly affect the CMR or the input impedance, and their value can, therefore, be chosen for the optimum frequency and offset voltage characteristics. Since the output differential amplifier is buffered by A_1 and A_2, its feedback resistors R_2 can be made low, providing the optimum CMR with frequency and minimum offset due to bias currents. The three operational amplifier instrumentation amplifiers can have very low equivalent input offsets if A_1 and A_2 are selected for matched offset voltage and offset voltage drift (modular amplifiers with drift as low as 0.25 μV/°C are available). The independent action of the gain-setting resistor allows gain ranges of 1 to 1000 V/V with less than 0.01 percent gain nonlinearity while maintaining very high CMR over the entire gain range. Typically the unity-gain CMR at 60 Hz with 1 kΩ source unbalance is 74 dB, and at a gain of 1000 it is 100 dB.

The circuit configuration of a typical low-cost FET input instrumentation amplifier is shown in Fig. 1.37b. This type provides very high input impedance, 10^{11} Ω differential and common-mode, with commensurately low input bias currents, 10 pA. These characteristics minimize the source loading and degradation of the CMR due to unbalanced source impedances. If a high-speed output operational amplifier is used, the design shown in Fig. 1.37b can also provide wideband, fast settling characteristics. Such performance is ideal in high-speed data acquisition systems, where a fast transient signal riding on a common-mode signal must be accurately amplified.

REFERENCES

1. G. E. Tobey, J. G. Graeme, and L. P. Huelsman, *Operational Amplifiers: Design and Applications*, McGraw-Hill Book Company, New York, 1971.
2. J. G. Graeme, *Applications of Operational Amplifiers: Third-Generation Techniques*, McGraw-Hill Book Company, New York, 1973.
3. J. E. Solomon, The Monolithic Op Amp: A Tutorial Study, *IEEE J. Solid-State Cir.*, December 1974.
4. D. Sheingold, Calculating Resistances for Sum and Difference Networks, *Electronics*, June 12, 1975.
5. W. E. Ott, Minimize Noise Problems with Instrumentation Amplifiers, *Electron. Products*, June 17, 1974.

2
LOGARITHMIC CONVERSION

The silicon bipolar transistor can be used as a very predictable nonlinear element. Its base-emitter voltage is a logarithmic function of collector current for currents from as low as few picoamperes to more than a milliampere. By connecting the transistor as a feedback element of an operational amplifier, accurate logarithmic conversion can be performed.[1] Logarithmic conversion provides a simple basis for performing many useful functions, including multiplication, division, general exponentiation, signal compression, and expansion. These functions can be combined with amplifiers to perform more complex operations, such as rms-to-dc conversion, vector addition, and power series expansion of virtually any function. It is the purpose of this chapter to provide the basic theory of logarithmic conversion, with an emphasis on such practical considerations of the circuit design as log conformity, frequency stabilization, and temperature effects.

2.1 Bipolar Transistor as a Fundamental Logarithmic Element

The collector current of the idealized intrinsic transistor was shown by Ebers and Moll[2] to be

$$i_C = \alpha_F I_{ES} (e^{qv_{BE}/kT} - 1) - I_{CS}(e^{-qv_{CB}/kT} - 1)$$

where α_F = forward current-transfer ratio
I_{ES}, I_{CS} = emitter and collector saturation current
v_{BE}, v_{CB} = base-emitter and collector-base voltages
q = charge of electron = 1 eV
k = Boltzmann's constant = 8.62×10^{-5} eV/°K
T = absolute temperature

For the forward-active operation, $i_C \gg \alpha_F I_{ES}$; with zero collector-base voltage, the base-emitter voltage is found to be

$$v_{BE} = \frac{kT}{q} \ln \frac{i_C}{I_S} \qquad (2\text{-}1)$$

where $\alpha_F I_{ES}$ has been shortened to simply I_S. The factor α_F should not be confused with the commonly used common-base current gain $\alpha = I_C/I_E$, since $\alpha = I_C/I_E$ includes voltage- and temperature-sensitive emitter-base leakage currents, and is always less than the relatively constant α_F. The major defect of these expressions is that they ignore the extrinsic emitter, base, and collector resistances (r_{ES}, r_{BS}, and r_{CS}, respectively) and the Early effect[3] (base-width modulation).

The emitter and base resistances cause the terminal base-emitter voltage to have a component which increases approximately linearly with increasing collector current. The effect of these two resistances can be adequately accounted for by a single resistance in the emitter of the logging transistor as shown in Fig. 2.1, where r_B is the total effective bulk resistance. The collector resistance r_{CS} causes the collector-base voltage of the intrinsic transistor to vary with collector current. The voltage drop on the collector resistance tends to forward-bias the intrinsic collector-base junction, producing an error current $I_{CS}\,(e^{-qr_{CS}/kT} - 1)$ in the collector current. Through the use of the power series expansion for e^x,

$$e^x = 1 + x + \frac{x^2}{2!} + \frac{x^3}{3!} + \cdots$$

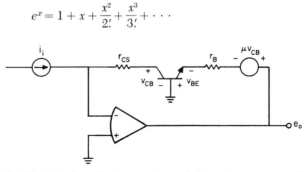

Fig. 2.1 Basic logarithmic converter, including primary sources of nonlogarithmic response.

it is found that for small error it is necessary that this forward bias be small compared with kT/q, in which case the base-emitter voltage is found to be

$$v_{BE} = \frac{kT}{q} \ln \left[\frac{i_C}{I_S} \left(1 + \frac{qI_{CS}r_{CS}}{kT} \right) \right]$$

Variation of the base-emitter voltage due to the collector-base bias is not explicit in the Ebers-Moll equation. However, a voltage feedback factor μ exists because the base width is modulated by the collector bias:

$$\mu = \frac{\delta v_{BE}}{\delta v_{CB}}$$

This phenomenon, first pointed out by Early[3] and commonly referred to as the Early effect, is accounted for in Fig. 2.1 by the voltage source μv_{CB}. For the circuit configuration in Fig. 2.1, where the terminal collector-base voltage is essentially zero, the only collector-base voltage variation is again due to the collector resistance r_{CS}. The voltage drop on the collector resistance produces a change in the terminal base-emitter voltage of $\mu i_C r_{CS}$.

The total base-emitter voltage, the output of the circuit in Fig. 2.1, is the sum of the intrinsic base-emitter voltage, the voltage drop on the bulk resistance r_B and the feedback voltage $\mu i_C r_{CS}$:

$$e_o = -\frac{kT}{q} \ln \left[\frac{i_i}{I_S} \left(1 + \frac{qI_{CS}r_{CS}}{kT} \right) \right] - i_i (r_B + \mu r_{CS})$$

For modern silicon transistors the collector saturation current is only about 0.1 pA and the collector saturation resistance is of the order of 5 to 100 Ω, which results in the term $(1 + qI_{CS}r_{CS}/kT)$ being different from 1 by less than 10^{-9}. The effective bulk resistance r_B ranges between 0.25 and 10 Ω, depending on the size of the transistor, and is generally large compared to μr_{CS}, since the feedback factor is typically 3×10^{-4} and as a result μr_{CS} is in the range of only 0.0015 to 0.03 Ω. The conclusion is that the only significant error in the simple expression for the base-emitter voltage given in Eq. (2-1) is the effect of bulk resistance in the emitter and base, provided the voltage drop on the collector saturation resistance as previously discussed is less than kT/q, which is about 26 mV at $+25°C$.

2.1.1 Compensation of the bulk resistance error   By applying a voltage of the proper magnitude to the base of the logging transistors as shown in Fig. 2.2, the error due to the voltage drop on the emitter and base bulk re-

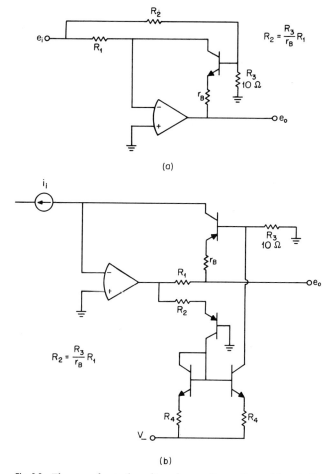

Fig. 2.2 The error due to the voltage drop on the emitter and base bulk resistances can be compensated by applying an equal and opposite voltage to the logging transistor's base.

sistances can be compensated[4,5]. The value of the compensation resistor R_3 should be kept small, since it increases the effective value of r_B by the amount R_3/β, where β is the common-emitter current gain of the transistor. The results of compensating the bulk resistance error are shown in Fig. 2.3. The uncompensated log conformity is 3 percent, and by applying the proper compensating voltage to the base of the logging transistor the log conformity was improved to better than 0.3 percent. The effective value of the bulk resistance for this transistor is 1.2 Ω.

50 Function Circuits

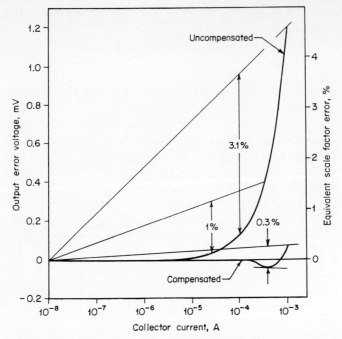

Fig. 2.3 Compensation of the junction bulk resistances by the circuits in Fig. 2.2 can improve the log conformity by a factor of 10 or more.

2.1.2 Transistor connection versus diode connection Perhaps the most useful connection of the logging transistor is as a diode, since it can easily be reversed to accommodate either polarity of input and can readily be stacked to perform more complex functions. Figure 2.4 shows a basic logarithmic converter using a diode-connected transistor, including compensating circuitry for the bulk resistance. This compensation network does introduce a small gain error, since the voltage drop on the

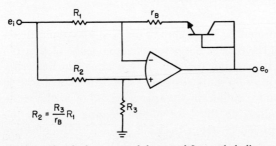

Fig. 2.4 The diode-connected log amplifier with bulk resistance compensation is particularly useful, since the diode-connected logging transistor can be reversed to accommodate the opposite-polarity input.

summing resistor R_1 is reduced by $e_i R_3/(R_2 + R_3)$. For a typical application where e_i has a maximum value of 10 V and the corresponding current is 1 mA, the gain error would only be 0.02 percent if the bulk resistance is 2 Ω, and this error is easily compensated by making R_1 0.02 percent low if it is significant.

The major deficiency of this connection is that not all the input current flows through the collector because of the collector-base connection. Experimental measurements of the collector and base current of an npn transistor are shown in Fig. 2.5 as a function of base-emitter forward bias. From these curves it can be seen that the collector current follows the exponential voltage dependence predicted by Eq. (2-1) over more than 8 decades. However, the base current is seen to have a different dependence, particularly at low currents. This results because for low levels

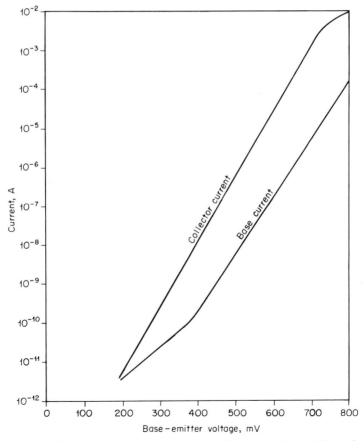

Fig. 2.5 Collector and base current as a function of base-emitter forward bias with zero collector-base voltage.

the base current is primarily the result of recombination in the base-emitter space-charge region.[6,7] The resulting dynamic range of the diode-connected logging transistor is only about 6 decades.

2.2 Logarithmic Amplifiers

The circuits described in the previous section represent basic logarithmic converters, but practical logarithmic amplifiers require further circuitry for compensating the effects of temperature, providing phase compensation, and protecting the base-emitter junction. The particular requirements on the operational amplifier used and the circuits for performing antilog and log ratios will also be discussed.

2.2.1 Phase compensation of log amplifiers

The log amplifiers shown in Figs. 2.1 and 2.2 tend to be unstable, since they have active gain inside the feedback loop. A simple way to analyze the frequency stability of these amplifiers is to open the feedback loop at the inverting input of the operational amplifier and then consider the frequency response of the two-stage amplifier.[8] Figure 2.6 shows the open-loop log amplifier along with its equivalent circuit, where C_c is the sum of the transistor's collector-base capacitance and the operational amplifier's open-loop input capacitance, r_e is the small-signal impedance of the emitter, and μ is the Early effect feedback factor discussed in Sec. 2.1. Phase compensation components R_2 and C_1 are also included.

The overall open-loop gain is

$$A = \frac{e_f}{e_d} = -A_1 \frac{e_f}{v_1}$$

In the complex frequency domain s, the open-loop output voltage is

$$E_f(s) = -(I_e + I_i) \frac{R_1}{1 + sC_cR_1}$$

and the emitter current is

$$I_e(s) = \frac{\mu E_f - V_1}{R_2 + r_e}$$

while the compensation capacitor current is

$$I_1(s) = (E_f - V_1) sC_1$$

and

$$I_2(s) = E_f \left(\frac{1}{R_1} + sC_c \right)$$

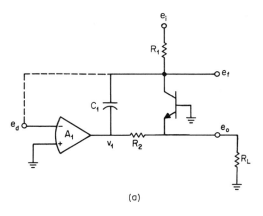

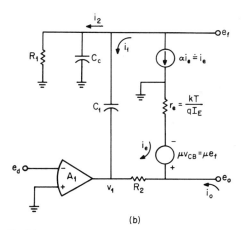

Fig. 2.6 Opening the feedback loop of the log amplifier simplifies the phase compensation analysis.

Combining these gives

$$\frac{E_f}{V_1} = \frac{R_1}{R_2 + r_e + \mu R_1} \frac{1 + sC_1(R_2 + r_e)}{1 + s[(C_c + C_1)(R_1)(R_2 + r_e)/(R_2 + r_e + \mu R_1)]}$$

For the sake of analysis it will be assumed that A_1 has a single-pole response (see Chapter 1, Sec. 1.3.1).

$$A_1(s) = \frac{A_{01}}{1 + s(A_{01}/2\pi f_T)}$$

where A_{01} is the dc open-loop gain and f_T the gain-bandwidth product of the operational amplifier. Now the overall open-loop gain can be written

54 Function Circuits

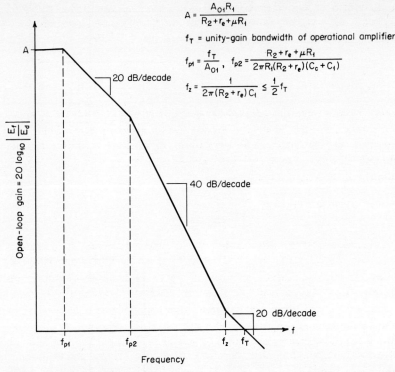

Fig. 2.7 Bode plot of open-loop log amplifier.

$$A(s) = \frac{-A_{01}R_1}{R_2 + r_e + \mu R_1}$$

$$\frac{1 + sC_1(R_2 + r_e)}{[1 + s(A_{01}/2\pi f_T)]\{1 + s[(C_c + C_1)(R_1)(R_2 + r_e)/R_2 + r_e + \mu R_1]\}}$$

A Bode plot of the log amplifier's open-loop gain is shown in Fig. 2.7. In order to ensure stability, the open-loop zero f_z should be made to occur at least 1 octave before the operational amplifier's unity-gain bandwidth, or $f_z \leq f_T/2$. This stability requirement ensures that the phase shift is less than 180° at the zero gain crossing, and the required phase compensation capacitance is then

$$f_z = \frac{1}{2\pi C_1(R_2 + r_e)} \leq \frac{1}{2}f_T \tag{2-2}$$

$$C_1 = \frac{1}{\pi(R_2 + r_e)f_T}$$

The amount of phase compensation is found to be a function not only of the phase compensation resistor, but also of the dc signal current, since the small-signal emitter resistance r_e is

$$r_e = \frac{kT}{qI_E}$$

In order to minimize the effects of the dc signal level on the phase compensation, R_2 should be made the maximum value the operational amplifier can accommodate. The maximum value of R_2 is limited by the maximum desired signal and load currents and the maximum output capability of the operational amplifier:

$$R_2 \leq \frac{V_{1\max} - 0.7}{(I_E + I_o)_{\max}}$$

As a practical example, consider the case of a typical operational amplifier with a 10 V output range and 1 MHz unity-gain bandwidth, and suppose the desired full-scale signal current and maximum load current are 1 mA each; then $R_2 \leq 4.65$ kΩ, and a practical value of 4.3 kΩ could be used, resulting in a minimum phase compensation capacitance of 74 pF.

The dynamic characteristics of the log amplifier output e_o are limited by the amount of phase compensation required. From Eq. (2-2), the -3 dB bandwidth of the logging transistor's collector current is the frequency of the open-loop response zero f_z, and since the zero frequency is a function of the dc signal level, the -3 dB bandwidth of the log conversion will be a minimum at the minimum input signal level. Perhaps more importantly, the step response of the log conversion will be nonlinear, having a slower fall time than rise time. This can be seen by inspection of Eq. (2-2) for the above example with both a signal current stepping from 1 nA to 1 mA and a signal current stepping from 1 mA to 1 nA. For the increasing step the time constant is continuously becoming shorter, approaching a final value of about 0.32 μs, while for the decreasing step the time constant is continuously lengthening, approaching a final value of about 1.9 ms.

2.2.2 Protection circuitry for the logging transistor

The logging circuits discussed are unipolar, and the operational amplifier has no dc feedback for the reverse input state. This means if the input is reversed, the amplifier's output will go to its lock-up state, which would reverse bias and possibly cause avalanche breakdown of the logging transistor's base-emitter junction. Avalanche breakdown of the logging transistor's base-emitter junction could also occur during the turn-on transient, since frequently one power-supply voltage will come up before the other, with the operational amplifier going into a saturation state. Avalanche breakdown of the base-emitter junction can result in a degradation of the transistor's gain and cause a significant increase in its noise. If the avalanche breakdown is maintained for a sufficient time, the junction will fail, while a reverse current of only 10 mA for a few seconds will typically increase the current noise several times and reduce the current gain considerably.[9]

A diode connected between the emitter and collector of the logging transistor as shown in Fig. 2.8a will provide dc feedback for either polarity of input, thereby preventing avalanche breakdown of the base-emitter junction. However, this approach introduces an error, particularly for low signal current levels, since the reverse leakage current of the protection diode subtracts from the input current. This error due to the protection diode's reverse leakage could be minimized by using a low-leakage FET with its drain and source tied together to form a low-leakage diode.

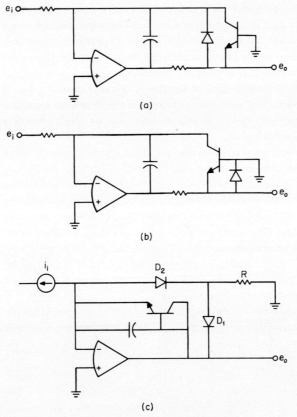

Fig. 2.8 Protection circuitry for the logging transistor.

Another simple approach is to connect the protection diode across only the base-emitter junction, as in Fig. 2.8b. In this manner the logging transistor's base-emitter junction is prevented from being reverse-biased sufficiently to avalanche, and the protection diode's reverse leakage produces no error since it represents only an insignificant load on the circuit's output. This approach has the drawback that the operational amplifier still has no dc feedback with reverse inputs and its output still goes to a saturation state. The major problem with the amplifier's output saturating is that the overload recovery time may be long, particularly if a chopper-stabilized amplifier is used.

The approach in Fig. 2.8b is also inadequate if the logging transistor is diode-connected, since the protection diode's reverse leakage will again produce an error. By the use of two protection diodes as in Fig. 2.8c, the logging transistor can be protected without introducing any significant errors while still providing dc feedback for the reverse input, and this configuration works equally well for either a transistor-connected or diode-connected logging transistor. For reverse inputs the protection diodes D_1 and D_2 in Fig. 2.8c are forward-biased and limit the reverse bias on the logging transistor's base-emitter voltage to about 1.4 V, which is generally insufficient to produce avalanche breakdown, thereby protecting the logging transistor as desired. During normal operation the protection diode D_1 is reverse-biased by the logging transistor's base-emitter voltage; however, its reverse leakage current flows through the resistor R, producing only a very small reverse bias on the protection diode D_2, whose reverse leakage would potentially produce an error in the log conversion. Since the reverse bias on D_2 can be kept very low, its leakage current and error contribution will be negligible.

2.2.3 Temperature effects in log amplifiers

The gain on the log conversion by a base-emitter junction is a linear function of temperature because of the kT/q multiplier. Also, the saturation current is a very strong function of temperature and is approximately[6,10]

$$I_S = BT^3 e^{-qV_{g0}/kT} \qquad (2\text{--}3)$$

where V_{g0} = bandgap voltage at $0°K$ = 1.11 V
B = temperature-independent constant related to doping levels and junction geometry

Figure 2.9 shows a circuit which uses matched logging transistors to cancel the effects of the saturation current. The voltage at the noninverting input A_2 is

$$v_1 = -\frac{kT}{q} \ln \frac{e_i}{R_1 I_{S1}} + \frac{kT}{q} \ln \frac{I_{REF}}{I_{S2}}$$

58 Function Circuits

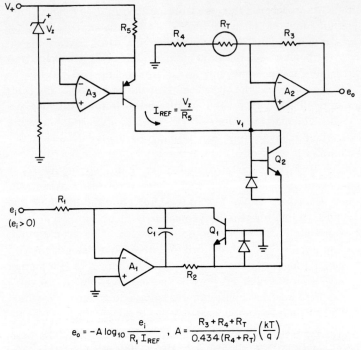

$$e_o = -A \log_{10} \frac{e_i}{R_1 I_{REF}}, \quad A = \frac{R_3 + R_4 + R_T}{0.434(R_4 + R_T)} \left(\frac{kT}{q}\right)$$

Fig. 2.9 A complete log amplifier using matched logging transistors Q_1 and Q_2 to cancel the effects of saturation current and a gain stage with a negative temperature coefficient to compensate for the positive temperature coefficient of the log conversion by a base-emitter junction.

which when combined with Eq. (2–3) simplifies to

$$v_1 = -\frac{kT}{q} \ln\left(\frac{e_i}{R_1 I_{REF}} \frac{B_2}{B_1}\right)$$

Note that this expression is essentially identical to Eq. (2–1) except that the uncontrollable saturation current has been replaced by an easily controlled reference current. Mismatch in the forward drop of the two logging transistors is due to $B_2/B_1 \neq 1$, and this does produce some error, but it is essentially temperature-independent and easily compensated by adjustment of the reference current. For circuit simplicity the diode connection of Q_2 has been used; this produces little error since Q_2 is used only as a reference and exact log conformity is unnecessary.

Gain temperature drift compensation has also been included in the log amplifier shown in Fig. 2.9. A thermistor R_T is used to produce a negative temperature coefficient in the output amplifier. The amplifier's gain would ideally vary inversely with temperature, but a temperature coefficient of about -3300 ppm/°C is generally sufficient.

$$v_1(T) \doteq v_1(T_0) + \left.\frac{\delta v_1}{\delta T}\right|_{T_0} (T - T_0)$$

and

$$\left.\frac{\delta v_1}{\delta T}\right|_{T_0} = \frac{v_1}{T_0} = +3.3 \times 10^{-3} v_1/°C$$

The gain of the output amplifier is

$$\frac{e_o}{v_1} = 1 + \frac{R_3}{R_4 + R_T}$$

and if R_3 and R_4 are low-drift resistors, the gain drift can be found by differentiation:

$$\frac{\delta(e_o/v_1)}{\delta T} = -\frac{e_o}{v_1} \frac{1}{R_4 + R_T} \frac{\delta R_T}{\delta T} \quad \text{for } R_3 \gg R_4 + R_T$$

A typical thermistor could have a positive temperature coefficient of about 7000 ppm/°C, which requires R_4 to have about the same resistance as the thermistor.

2.2.4 Requirements on the operational amplifier

The operational amplifier in a log converter is frequently the primary source of error. Its input bias current and input offset voltage subtract directly from the input current or voltage, and its bandwidth determines the upper limit on the log-conversion bandwidth.

The input bias current and offset voltage error sources can be modeled as in Fig. 2.10. For a current-source input the operational amplifier's offset voltage produces no error in the input current, and for reasonable offset (<26 mV) the bias on the collector-base junction will produce a constant error of only $(qV_{os}/kT)I_{CS}$, which is typically less than 0.1 pA (this is derived from a series expansion of the second term of the Ebers-Moll equation given in Sec. 2.1). Therefore, only the error due to the

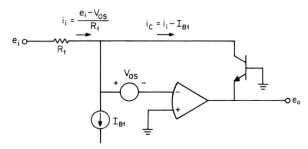

Fig. 2.10 Basic log amplifier, including the input bias current and offset voltage error sources.

input bias current need be compensated, and this can be easily done by supplying a high-impedance divider from supply to the summing junction as shown in Fig. 2.11a. The primary limitation of the bias current compensation is that the bias current noise and drift with temperature remain. The bias current drift with temperature of the typical bipolar input operational amplifier is about -1.0% °C, and a bias current of the order of 50 nA, which yields a drift of about 0.5 nA/°C, is not uncommon. Even in the laboratory this would limit the compensation to about ± 1 nA.

A simple FET buffer utilizing a matched pair of FETs can be built, as in Fig. 2.11b,[5] which will reduce the input bias current to about 10 pA. FET Q_1 serves as a source follower with essentially unity gain; it is biased with a matching FET current source Q_2. The drain current of Q_2 is V_{GS2}/R, and the offset due to the buffer is then $V_{GS1} - V_{GS2}$.

With a voltage input the bias current of the operational amplifier can be cancelled by inserting a resistor equal to the summing resistor R_1 in

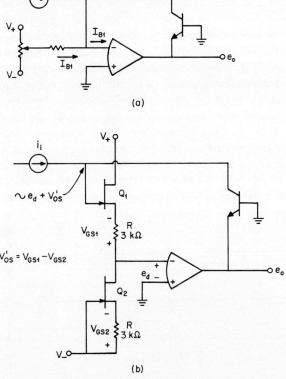

Fig. 2.11 The operational amplifier's input bias current can be compensated as in (a) or buffered with a FET source follower as in (b).

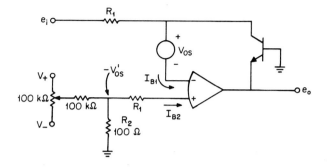

Total offset = $V'_{OS} \cong V_{OS} + R_1(I_{B1} - I_{B2})$

Fig. 2.12 A voltage input log amplifier can use balanced source impedance to cancel the operational amplifier's bias current, and the total effective offset can be compensated by a low-impedance voltage divider from supply.

series with the noninverting input of the amplifier, as shown in Fig. 2.12. The remaining input bias current error is only the input difference current times the value of the summing resistor, which is typically only one-tenth of the original bias current error. The effect of the remaining bias current error can be compensated along with the operational amplifier's input offset voltage. A simple means of compensating the total offset is also shown in Fig. 2.12, where a low-impedance voltage divider is used to offset the amplifier's noninverting input by a negative amount equal to the total effective offset. The resistor R_2 in the voltage divider must be kept small compared to R_1 for the bias current cancellation to be effective.

2.2.5 Antilog and log ratio amplifiers

If the logging transistors are connected in a suitable configuration, they can be used to perform accurate antilogging. A simple circuit is shown in Fig. 2.13, where the input voltage is attenuated on a thermistor and applied as a differential voltage to a pair of logging transistors:

$$v_{BE2} = V_{BE1} + \gamma e_i$$

where

$$\gamma = \frac{R_T}{R_3 + R_T}$$

The base-emitter voltage of logging transistor Q_1 is a reference to cancel the saturation current, and from Eq. (2-1),

$$\frac{kT}{q} \ln \frac{i_{C2}}{I_{S2}} = \frac{kT}{q} \ln \frac{I_{REF}}{I_{S1}} - \gamma e_i$$

62 Function Circuits

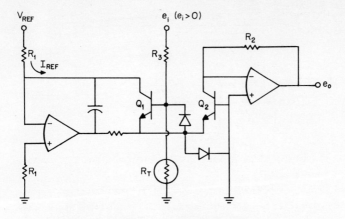

$$e_o = R_2 I_{REF} 10^{-e_i/A} \quad, \quad A = \frac{R_3 + R_T}{0.434 R_T}\left(\frac{kT}{q}\right)$$

Fig. 2.13 Antilog amplifier using matched logging transistors Q_1 and Q_2 to cancel the effects of saturation current and a thermistor to compensate the positive temperature coefficient of the base-emitter junction log converter.

which combined with Eq. (2–3) yields

$$e_o = \frac{B_2}{B_1} R_2 I_{REF} e^{\frac{-q\gamma}{kT} e_i}$$

For most applications $R_3 \gg R_T$, and if the thermistor drift is $+3300$ ppm/°C, the temperature drift will be approximately compensated as discussed in Sec. 2.2.3.

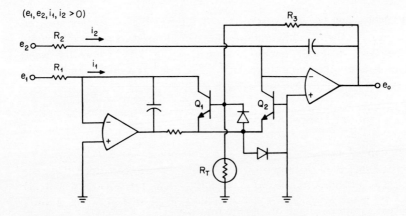

$$e_o = A \log_{10} \frac{i_1}{i_2} \quad, \quad A = \frac{R_3 + R_T}{0.434 R_T}\left(\frac{kT}{q}\right)$$

Fig. 2.14 Log ratio of either current or voltage inputs is easily performed with a pair of logging transistors.

A variation of the antilog amplifier discussed above is the log ratio circuit in Fig. 2.14. For this configuration the output is

$$e_o = \frac{kT}{q\gamma} \ln \left(\frac{i_1}{i_2} \frac{B_2}{B_1} \right)$$

where

$$\gamma = \frac{R_T}{R_3 + R_T}$$

as in the antilog circuit. The inputs can be either voltage or current sources with this log ratio circuit.

All the previous log amplifiers have used npn-type logging transistors, which results in the input necessarily being positive. If the application requires a negative input, the log amplifiers can be built using pnp-type logging transistors. An example of the pnp implementation of the log ratio circuit is shown in Fig. 2.15.

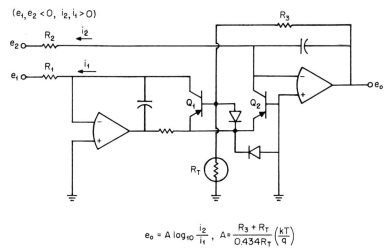

$$e_o = A \log_{10} \frac{i_2}{i_1}, \quad A = \frac{R_3 + R_T}{0.434 R_T} \left(\frac{kT}{q} \right)$$

Fig. 2.15 Negative inputs can be accommodated by using pnp-type logging transistors, as in this log ratio circuit.

2.2.6 Current inverters[11]

Because of processing limitations beyond the scope of this book to explain, the pnp transistor is generally unsuitable for use as a logging transistor in integrated-circuit form, and consequently most IC logging circuits utilize npn logging transistors. This requires that the input current be into the log amplifier if the transistor connection of the logging transistor is used. Simple current inverters, as shown in Fig. 2.16, can be inserted in series with the input to the log amplifier to accommodate the opposite polarity of the input signal. The straightforward current inverter in Fig. 2.16a uses resistors as the feedback elements of an operational amplifier to provide the transfer function

$$i_2 = \frac{R_1}{R_2}(i_1 + I_{B1} - I_{B2}) + \frac{V_{OS}}{R_2}$$

where V_{OS}, I_{B1}, and I_{B2} are the input offset voltage and bias currents of the amplifier. The input bias currents are typically matched and produce only a small error compared with the offset voltage. In typical applications the current gain is -1 and the full-scale output of the operational amplifier is ± 10 V, which results in the maximum dynamic signal range for 1 percent log conversion being only about 4 decades, even under laboratory conditions where the offset voltage can be held to less than about 10 μV.

If greater dynamic range is required, or particularly if the ambient

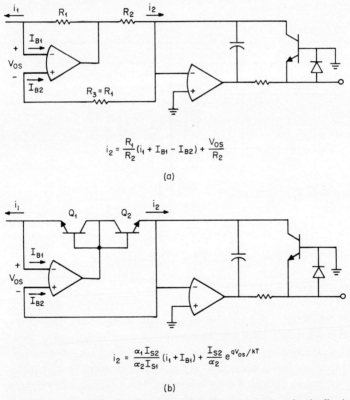

Fig. 2.16 A straightforward current inverter using resistors as the feedback elements of an operational amplifier only provides a dynamic range of about 4 decades, while a logarithmic current inverter using a pair of logging transistors provides 6 decades of dynamic range.

temperature of the operational amplifier cannot be so carefully controlled and offset voltages of the order of 100 μV or more must be tolerated, a logarithmic current inverter as shown in Fig. 2.16b should be used. With Eq. (2-1) the transfer function of the current inverter using logging transistors as feedback elements can be determined; it is

$$i_2 = \frac{\alpha_1 I_{S2}}{\alpha_2 I_{S1}}(i_1 + I_{B1}) + \frac{I_{S2}}{\alpha_2} e^{qV_{os}/kT}$$

Note that in this case the input bias currents do not cancel, even if they are matched, and will represent the major error if well-matched logging transistors, such as a monolithic pair, are used for the feedback elements. The saturation current I_S is of the order of 0.1 pA for modern silicon npn transistors, and α is nearly 1; therefore, from the power series expansion of e^x, if the offset voltage is less than $kT/q \doteq 26$ mV, the error due to the offset voltage will be of the order of I_S or only 0.1 to 0.2 pA. If an FET input amplifier is used and its input bias current is held to less than 10 pA, the dynamic range for 1 percent log conversion will be 6 decades for a 1 mA maximum current. There will be a scale-factor error because of mismatch between the current gains and the saturation currents of the logging transistors, and scale-factor drift versus temperature will also result because of imperfect tracking between these parameters of the two transistors. With a well-matched monolithic pair, the scale-factor error will typically be less than 4 percent and can be compensated for by a scale-factor adjustment in the log amplifier; the remaining scale-factor drift versus temperature would typically be less than 80 ppm/°C.

2.5 Log Amplifier Specifications and Testing

The ideal output of a log amplifier for either a current input i_i or voltage input e_i is

$$e_o = -A \log_{10} \frac{i_i}{I_{REF}} \qquad \text{current input}$$

$$e_o = -A \log_{10} \frac{e_i}{RI_{REF}} \qquad \text{voltage input}$$

The negative amplifier gain is not necessary or ideal as such; however, it is common, and to be consistent with the log amplifier described in Sec. 2.2.3 and shown in Fig. 2.9, it is assumed to be desired. The constants A and I_{REF} are each independently set.

To minimize the effects of output offset and noise, the full-scale output should usually be set at the maximum output specification of the opera-

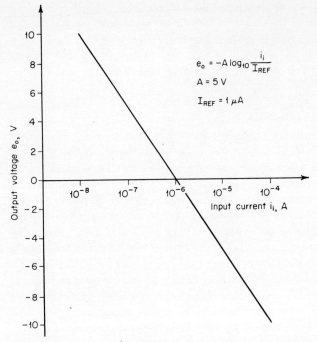

Fig. 2.17 Response of log amplifier.

tional amplifier, typically ± 10 V. For a full-scale output range of $\pm e_{\text{ofs}}$, an input range of $i_{\min}$ to $i_{\max}$ means

$$+e_{\text{ofs}} = -A \log \frac{i_{\min}}{I_{\text{REF}}} \quad \text{and} \quad -e_{\text{ofs}} = -A \log \frac{i_{\max}}{I_{\text{REF}}}$$

$$i_{\min} \leq i_i \leq i_{\max}$$

Adding these two equations yields

$$\log \frac{i_{\max} i_{\min}}{I_{\text{REF}}^2} = 0$$

so that

$$I_{\text{REF}} = \sqrt{i_{\max} i_{\min}}$$

And now the value of the scale factor A is found to be

$$A = \frac{e_{\text{ofs}}}{\log_{10} \sqrt{i_{\max}/i_{\min}}}$$

As an example, assume that the minimum input current is 10 nA and the maximum input current is 100 μA, with a ± 10 V full-scale output

$$I_{REF} = \sqrt{(10 \text{ nA})(100 \text{ }\mu\text{A})} = 1 \text{ }\mu\text{A}$$
$$A = \frac{10}{\log_{10} \sqrt{10^{-4}/10^{-8}}} = 5 \text{ V}$$

The transfer function of the log amplifier with these values of A and I_{REF} is plotted in Fig. 2.17.

2.3.1 Accuracy and dynamic range

Once A and I_{REF} have been set, the accuracy specification puts an upper limit on the amount of allowable deviation of the output voltage from the ideal log response. For example, the specification of ± 1 percent of full-scale accuracy means the output must be within ± 100 mV of the ideal value for a 10 V full-scale output. The accuracy specification puts a bound on the output error voltage ϵ_o. This error can be referred to the input by manipulating the transfer function:

$$\epsilon_o = -A \log \frac{i_i}{I_{REF}} - \left[-A \log \frac{i_i + \Delta i_i}{I_{REF}}\right]$$

$$\epsilon_o = A \log \left(1 + \frac{\Delta i_i}{i_i}\right) \quad \text{error referred to output}$$

$$\Delta i_i = i_i(10^{\epsilon_o/A} - 1) \quad \text{error referred to input}$$

where Δi_i is the error referred to the input. The primary sources of input error are the operational amplifier's input bias current and offset voltage. As discussed in Sec. 2.2.4, with current-source inputs the offset voltage has little effect, and the operational amplifier's input bias current can be reduced to less than 10 pA by using a FET input stage. For 1 percent accuracy the minimum input current would be 100×10 pA $= 1$ nA. If a voltage source is used, the amplifier's input offset will also introduce an error current of V_{OS}/R, where V_{OS} is the offset voltage and R the summing resistor. The offset can be nulled as shown in Fig. 2.12; however, even under laboratory conditions it can only be reduced to about 2°C of offset drift, which is typically 5 μV for bipolar input operational amplifiers and 20 μV for FET input amplifiers. The minimum input voltage for 1 percent accuracy is then 0.5 to 2 mV with typical bipolar and FET input operational amplifiers. At high current levels the bulk resistance of the emitter and base will produce an output error which is a fraction of the full-scale output. As discussed in Sec. 2.1.1, the bulk resistance will produce an error of the order of 1 percent of full scale without special compensation for currents up to about 300 μA.

2.3.2 Output voltage drift with temperature

It is typical of industrial environments that the errors due to ambient temperature variation are the dominant system errors. By compensating the scale-factor drift with a thermistor, cancelling the saturation current with a matching logging transistor as discussed in Sec. 2.2.3, and using low-drift operational amplifiers, the temperature-variation errors can be minimized. However, there will still be an output drift with temperature, and this drift can be approximated as

$$\frac{\Delta e_o}{\Delta T} = \frac{\Delta A}{A\,\Delta T} e_o + 0.434 A \left(\frac{\Delta I'_{REF}}{I'_{REF}\,\Delta T} + \frac{\Delta I_B}{i_i\,\Delta T} + \frac{\Delta V_{OS}}{R i_i\,\Delta T} \right)$$

where I'_{REF} is the effective reference current including the effects of mismatch between the saturation current of the two logging transistors, and $\Delta I_B/\Delta T$ and $\Delta V_{OS}/\Delta T$ are the operational amplifier's input bias current and offset voltage drift with temperature.

The scale-factor drift $\Delta A/A\,\Delta T$ is primarily due to imperfect compensation of the kT/q gain on the base-emitter junction log conversion. As shown in Sec. 2.2.3, the initial drift is about +3300 ppm/°C, and this can readily be compensated to less than ±400 ppm/°C through the use of a thermistor. The effective reference current drift $\Delta I'_{REF}/I'_{REF}\,\Delta T$ is due to the combination of the I_{REF} drift, which is easily held to less than 100 ppm/°C, and the residual differential drift of the saturation currents, which is typically less than 80 ppm/°C if a monolithic pair of transistors is used for the logging transistors.

2.3.3 Calibrating the log amplifier

The first step in calibrating the log amplifier is to null the input offset voltage of the operational amplifier if the input is a voltage source. It is generally not necessary to null the operational amplifier's offset voltage if a current-source input is used. An appropriate circuit for nulling the offset voltage is shown in Fig. 2.12, and this is easily performed by adjusting the reference current I_{REF} to its approximate desired value and supplying an input current approximately equal to I_{REF}. Then adjust the offset voltage for a minimum change in the output while switching the input of the summing resistor between being open-circuited and being grounded. The reference current is easily adjusted by applying an input current i_i equal to the desired value of I_{REF} and then adjusting the reference current for zero output voltage (adjust R_5 in Fig. 2.9).

$$e_o = -A \log_{10} \frac{i_i}{I_{REF}} = 0 \quad \text{for } i_i = I_{REF}$$

After I_{REF} has been set, apply the maximum input signal corresponding to i_{max} and adjust the output amplifier's gain (adjust R_3 in Fig. 2.9) for the proper full-scale output. If a current-source input is used, the operational amplifier's input current can now be nulled with the circuit in Fig. 2.11

by applying the minimum input current and adjusting the bias current compensation for the proper full-scale output.

2.3.4 Testing the log amplifier's dynamic performance

Log amplifiers are unipolar devices, and their bandwidth is generally specified for small-signal variations about a dc bias input. The bandwidth of the log amplifier is a function of the signal level due to the dynamic impedance of the emitter $r_e = kT/qI_E$, as discussed in Sec. 2.2.1. The small-signal bandwidth can be tested by summing a small ac input with the dc level of interest. The peak-to-peak value of the ac input should be about 5 percent of the dc level; its frequency can be increased until the alternating current at the output rolls off 3 dB. For the phase compensation discussed in Sec. 2.2.1, the -3 dB bandwidth for various current levels would typically be:

Input current	-3 dB bandwidth
1 mA	500 kHz
100 μA	470 kHz
10 μA	310 kHz
1 μA	71 kHz
100 nA	8.2 kHz
10 nA	830 Hz
1 nA	83 Hz

These bandwidths would be achieved at the noninverting input of the output amplifier, v_1, for the log amplifier in Fig. 2.9. However, because of the finite bandwidth of the output amplifier, the output voltage e_o may have a narrower bandwidth. If the scale factor $A = 5$ V and a typical IC output amplifier with a 1 MHz unity-gain bandwidth is used, the output stage will limit the overall bandwidth to about 100 kHz even with input currents greater than 1 μA.

The step response is also a function of the signal level, as discussed in Sec. 2.2.1, and is best measured by summing two inputs, one a dc level equal to the minimum current of interest and the other a unipolar square wave with a peak value providing the maximum current of interest. As discussed in Sec. 2.2.1, the step response will typically be unsymmetrical, with a longer fall time than rise time.

REFERENCES

1. J. F. Gibbons and H. S. Horn, A Circuit with Logarithmic Transfer Response over 9 Decades, *IEEE Trans. Circ. Theory,* September 1964.
2. J. J. Ebers and J. L. Moll, Large-Signal Behavior of Junction Transistors, *Proc. IRE,* December 1954.
3. J. M. Early, Effects of Space-Charge Layer Widening in Junction Transistors, *Proc. IRE,* November 1952.

4. D. R. Morgan, Get the Most out of Log Amplifiers by Understanding the Error Sources, *EDN*, Jan. 20, 1973.
5. J. G. Graeme, *Applications of Operational Amplifiers: Third-Generation Techniques*, McGraw-Hill Book Company, New York, 1973.
6. A. S. Grove, *Physics and Technology of Semiconductor Devices*, John Wiley & Sons, Inc., New York, 1967.
7. C. L. Indindoli, Surface State Charge Effects on Very Low Current Transistor Characteristics, *Proc. IEEE*, June 1970.
8. G. E. Tobey, J. G. Graeme, and L. P. Huelsman, *Operational Amplifiers: Design and Applications*, McGraw-Hill Book Company, New York, 1971.
9. C. D. Motchenbacher, Protect Your Transistors against Turn-on or Testing Transient Damage, *Electronics*, Dec. 6, 1971.
10. P. E. Gray, D. DeWitt, A. R. Boothroyd, and J. F. Gibbons, *Physical Electronics and Circuit Models of Transistors*, John Wiley & Sons, Inc., New York, 1964.
11. B. Gilbert, Current Inverter with Wide Dynamic Range, a Useful Adjunct to Integrated-Circuit Log Devices, *Analog Dialogue*, Analog Devices, Inc., vol. 9, no. 2, 1975.

3
MULTIPLIERS, DIVIDERS, AND MULTIFUNCTION CONVERTERS

Analog multipliers and dividers were originally used as building blocks for analog computers. Today they are still performing analog computations, but not so much in analog computers anymore. Instead, most are used as signal processors in flow computations, guidance systems, displays, chemical analyzers, photo processing equipment, motor controls, modulators, and a variety of other on-line applications. In manipulating signals derived from physical processes, it is often necessary to realize an output signal that is a nonlinear function of an input. Thus, there is a need for a circuit element whose nonlinear transfer function can be tailored arbitrarily. An analog function block called a *multifunction converter* has recently been developed. It will perform

$$e_o = v_y \left(\frac{v_z}{v_x}\right)^m$$

where m is adjustable by two external resistors, and v_x, v_y, and v_z are input voltage variables.

In recent years, integrated-circuit and hybrid technologies have already played and will continue to play an important role in the manufacture of analog circuit functions. The keys to successful design and production of such functions will be extremely close device matching and combined technologies, i.e., bipolar, FET, thin-film, and laser trimming. The performance of analog circuits will continue to improve as these better techniques of matching and trimming evolve and become more routine. And, as this happens, prices will continue to fall. Compared with those of the simplest digital circuits, the prices of these analog circuits will remain high because of the need for tight process controls and for trimming and comprehensive testing. However, on a "per function" basis, the analog approach will remain the lowest cost and most reliable approach to problem solving.

3.1 General Description and Multiplier Specifications

The ideal analog multiplier is essentially a simple device. It provides an output voltage (e_o) as a function of two voltage inputs (x,y) such that

$$e_o = \frac{xy}{k}$$

where k is an appropriate constant. Furthermore, an ideal multiplier will have infinite input impedance for x and y, and the output will be an ideal voltage source (zero output impedance). Naturally no practical multiplier meets any of these ideal requirements, so multiplier specifications are created to attempt to characterize reality as opposed to the ideal. Definition of multiplier specifications has provided a continual challenge to manufacturers' imagination, while understanding and applying the specifications can try the patience of almost any user. This section will shed some light on the area of multiplier specifications.

3.1.1 Simple departures from the ideal multiplier Practical multipliers have a number of basic limitations which serve to limit performance and operating ranges. Most multipliers are intended for use in analog systems with standard ±15 V power supplies. Hence the absolute maximum input signals are usually ±15 V, and for normal operation ±10 V are the typical limits for x and y. Often k is chosen so that the output signal range equals the input signal range, i.e.,

$$k = \sqrt{x_{max} y_{max}}$$

Thus most analog multipliers have $e_o = xy/10$, which provides ±10 V output for ±10 V inputs. In addition to having a finite output voltage range, all practical multipliers can provide only a limited output current, with 5 and 10 mA being fairly standard values. Virtually all present-day multipliers have output current limiting circuitry to protect the output stage from damage if the output is shorted to power-supply common. As long as the output is not in current limit, most multipliers have low, but not zero, output impedance. Practical multipliers also have finite input impedance; in fact, many multipliers have input impedance between 10 and 100 kΩ. Low input impedance can cause serious errors if the source impedance is not small. For example, a 0.1 percent accurate multiplier with 10 kΩ input impedance would have an additional 0.1 percent loading error for input source impedances of only 10 Ω.

The specifications considered so far are circuit limitations. The basic accuracy of the multiplier is not affected so long as external constraints are followed. We will call these specifications *dc circuit constraints*. In summary, they are

1. Rated output (minimum voltage at rated current)
2. Output impedance
3. Maximum input voltage
   (a) Maximum voltage for rated specifications
   (b) Maximum voltage for no damage to device
4. Input impedance

3.1.2 Multiplier dc error specifications

We now come to the heart of multiplier specifications—the accuracy and error specifications. These are the specifications which determine how well a multiplier performs under optimum external conditions. Multiplier inaccuracies have two basic sources—offsets and nonlinearities. The multiplier is inherently a nonlinear device, but we will ultimately attempt to define "linearity." For the present we will consider operation at direct current only. To facilitate our discussion, consider the following algebraic model for a multiplier:

$$e_o = (1 + k_e)\frac{(x + X_{os})(y + Y_{os})}{10} + V_{os} + n(x,y) \qquad (3\text{--}1)$$

where
e_o = output voltage
k_e = gain error
x, y = input voltages
X_{os} = offset voltage associated with the x input terminal
Y_{os} = offset voltage associated with the y input terminal
V_{os} = output offset voltage
$n(x,y)$ = nonlinearities, terms such as x^2, y^2, x^2y, xy^2, etc. By definition, we assume $n(0,0) = 0$

We can simplify Eq. (3–1) for e_o by multiplying and eliminating second- or higher-order terms ($X_{os}Y_{os}$, k_eX_{os}, etc.). Then

$$e_o = \underbrace{\frac{xy}{10}}_{\substack{\text{Ideal}\\\text{output}}} + \underbrace{\frac{k_e xy}{10}}_{\substack{\text{Gain}\\\text{error}}} + \underbrace{\frac{xY_{os}}{10}}_{\substack{y\text{ input}\\\text{offset}}} + \underbrace{\frac{X_{os}y}{10}}_{\substack{x\text{ input}\\\text{offset}}} + \underbrace{V_{os}}_{\substack{\text{Output}\\\text{offset}}} + \underbrace{n(x,y)}_{\text{Nonlinearity}} \qquad (3\text{--}2)$$

Total error

Keeping Eq. (3-2) in mind, we will proceed to define the various multiplier error terms.

Total error (accuracy). Total error is the actual departure of the multiplier output voltage e_o from the ideal product $xy/10$. The total error may be specified in millivolts or in percent of full scale (normally 10 V). Most manufacturers specify a maximum value for total error. To arrive at the maximum value, the multiplier is tested at several different values of x and y, including points in all four quadrants. The multiplier must be within its maximum total error specification at all points.

INDIVIDUAL ERRORS

In general the total error specification is the best measure of multiplier quality, but certain applications may place a premium on certain components of the total error. For example, if x and y were always large values, offset errors would not be significant but gain accuracy would be critical.

Output offset. The output offset, V_{os} in Eq. (3-2), is defined as the output voltage when both inputs, x and y, are grounded. Output offset can be nulled in most multipliers. Like the offset voltage of operational amplifiers, V_{os} is sensitive to temperature and power-supply variations.

Input offset (feedthrough). An ideal multiplier has zero output when either of its inputs is at zero. Practical multipliers have a nonzero output when a voltage is applied to one input and the other input is at zero. This output is called *feedthrough*. Feedthrough is specified as the peak-to-peak output voltage when one input equals zero and the other input varies between +10 and −10 V. Examination of Eq. (3-2) shows that part of the feedthrough results from the input offset terms, but nonlinearity may have a greater effect on feedthrough. Input offsets are seldom specified separately, since they contribute only to feedthrough. The input offsets may often be trimmed, thus reducing feedthrough somewhat, but the effect of nonlinearity on feedthrough for a sine-wave input normally has a predominant second-harmonic content. The amount of signal feedthrough is usually different for the x and y inputs, and separate specifications for x and y feedthrough are common. Feedthrough is also strongly dependent on input frequency and is sometimes referred to as *ac null suppression*. Performance curves usually show the variation of feedthrough with frequency. Feedthrough or ac null suppression is usually most critical in such applications as product modulation or demodulation. Specification table numbers for feedthrough represent the dc or low-frequency feedthrough characteristics.

Gain error. Conceptually gain error and nonlinearity are separate concepts, but in practice they are rather hard to distinguish clearly. We have arbitrarily defined gain error as the output error at $x = y = +10$ V, for both output offset and input offsets nulled.

Linearity. Defining linearity for a device which generates a nonlinear function of two variables sounds like a contradiction. However, what is usually done is to fix one input. Then e_o is a "linear" function of the other input, e.g.,

$$e_o\bigg|_{y=10} = x \quad \text{or} \quad e_o\bigg|_{x=-1} = -\frac{y}{10}$$

Thus for any fixed value of voltage applied to one multiplier input, we can determine output linearity in a classical sense. In virtually all cases output nonlinearity is greatest when the constant input is at $+10$ or -10 V. Thus maximum x nonlinearity is determined as follows:

1. Set $y = +10$ V.
2. Determine e_o versus x for -10 V $\leq x \leq +10$ V.
3. Maximum nonlinearity is the maximum departure from a best-fit straight line for the function $e_o(x)|_{y=+10}$, as shown in Fig. 3.1.
4. Repeat steps 1, 2, and 3 for $y = -10$ V.
5. Maximum x nonlinearity is the greater of the values found for $y = +10$ V and $y = -10$ V.

In a similar fashion, y nonlinearity is determined by letting $x = +10$ V and -10 V, respectively.

A word of caution. We have now covered all the fundamental static (dc) specifications. Remember that all the parameters will be a function of temperature and power-supply variations. Not all manufacturers specify both temperature-drift and power-supply sensitivity for total error and for the individual major error terms.

Remember, too, that the total error specification includes all the in-

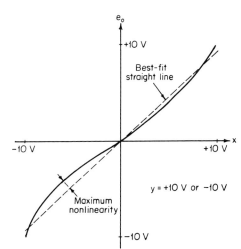

Fig. 3.1 Definition of multiplier x nonlinearity.

dividual terms [see Eq. (3-2)]. Maximum total error will always be less than the sum of the maximum individual errors, because the individual errors do not have their maximums at the same x, y operating point.

3.1.3 Dynamic performance
So far our discussions have only involved dc characteristics. Most of the parameters change with signal frequency, but usually only a few important parameters are specified over a frequency range. The parameters of interest are accuracy, feedthrough, and non-linearity or distortion.

Multiplier accuracy versus frequency is normally defined differently from dc accuracy. If x and y are two independent ac inputs, the complexity of the possible outputs is overwhelming. The problem is further compounded if x and y are nonsinusoidal signals. To simplify ac error analysis, one multiplier input is held at a fixed dc value and a sinusoidal signal is applied to the other input. Then various ac output errors may be characterized in a "normal" fashion.

The limited ac bandwidth of the multiplier is usually the most significant factor in ac accuracy. The ac bandwidth is most simply and commonly described through a small-signal Bode plot (magnitude and phase versus frequency response plots).

As indicated by the Bode plot, the multipliers have two basic types of frequency-related errors—magnitude errors and phase shift. If the multiplier is used as part of a closed-loop system, the phase shift which corresponds to loop delay may cause serious stability or accuracy problems. In "open-loop" systems, the multiplier magnitude error usually is the only significant problem.

Multiplier magnitude errors are easily measured using an ac voltmeter and/or an oscilloscope. The following magnitude versus frequency characteristics are usually specified by most manufacturers.

Small-signal frequency response. With one input at $+10$ or -10 V dc and the other input a small ac signal, normally 1 V p-p, the output -3 dB frequency is measured. A small-signal Bode plot is usually included for reference purposes.

Full-power frequency response. Full-power frequency response is the maximum frequency at which the output will swing 20 V p-p into a rated load. To measure this, one input is set at $+10$ or -10 V dc, and the other input is connected to a sine-wave generator whose amplitude is continuously adjusted to swing 20 V p-p at the multiplier output.

Bandwidth for dc accuracy. Bandwidth for dc accuracy is the maximum frequency for which the multiplier ac amplitude error is less than the dc error specified. This is measured like the small-signal frequency response, only here we require $e_o = e_i \pm$ dc accuracy. Occasionally, instead of specifying dc accuracy bandwidth, the frequency for 1 percent magnitude error is specified.

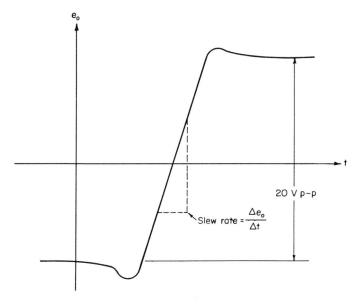

Fig. 3.2 Slew-rate test output signal. The input is a ± 10 V square wave, normally at 1 kHz.

Output slew rate. Output slew rate is measured with one input at $+10$ or -10 V and the other input a ± 10 V square wave. Slew rate then represents the maximum de_o/dt at the multiplier output waveform. To measure it, $\Delta e_o/\Delta t$ is used, as illustrated in Fig. 3.2. The slew rate and full-power frequency response are closely related.

Settling time. Settling time is the time required for the output to respond to a step input and to settle within some specified error band around the output final value. Settling time includes delay time, rise time, and ringing, as illustrated in Fig. 3.3a. For simplicity, settling time is usually tested with one input at -10 V and the other input connected to a ± 10 V square-wave generator, as in Fig. 3.3b.

Phase shift. Phase shift may be measured directly using an oscilloscope and is normally included as part of the Bode plot. However, small phase-shift angles are very difficult to measure directly on an oscilloscope but may, nevertheless, cause serious problems in closed-loop systems. If a precision phase-angle meter is not available, the concept of vector error may be used to better describe and measure very small amounts of phase shift. Figure 3.4a illustrates what is meant by vector error. Applying the trigonometric identity

$$\sin A - \sin B = 2 \cos \frac{1}{2}(A+B) \sin \frac{1}{2}(A-B)$$

to e_i and e_o, we have

$$e_i - e_o = K \sin wt - K \sin (wt - \phi)$$
$$= 2K \sin \frac{\phi}{2} \cos \left(wt - \frac{\phi}{2}\right)$$

Using the approximation $\theta \doteq \sin \theta$ for small θ, $(e_i - e_o)$ becomes

$$e_i - e_o = \phi K \cos \left(wt - \frac{\phi}{2}\right)$$

Then $(e_i - e_o)$ can be measured directly using a differential ac voltmeter, as in Fig. 3.4b. The reading on the voltmeter is directly proportional to the angle difference ϕ in radians. As an example, for $e_i = e_o = 7.07$ V rms, a reading of 70.7 mV rms on the differential ac voltmeter indicates 1 per-

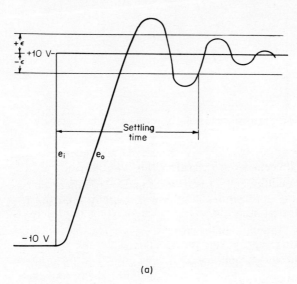

(a)

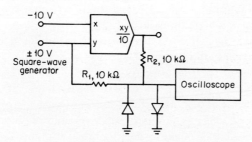

(b)

Fig. 3.3 (a) Illustration of settling time. (b) Test circuit for multiplier settling time.

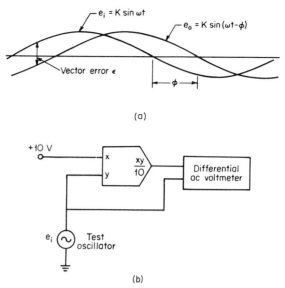

Fig. 3.4 (a) Illustration of vector error and phase shift. (b) Vector error measurement circuit.

cent vector error, or $\phi = 0.01$ rad (0.573°). Rather than trying to measure $\tfrac{1}{2}°$ phase shift directly, vector error measurements are commonly used.

Ac feedthrough. Feedthrough is usually strongly dependent on frequency. In modulation applications the term *ac null suppression* is a more apt description of feedthrough. Feedthrough versus frequency is tested with one input grounded and a 20 V p-p sine wave on the other. Feedthrough versus frequency should always be specified for each input, since there may be significant differences between the x and y channels with respect to this characteristic.

Nonlinearity versus frequency. In ac applications, multiplier nonlinearities add harmonic distortion. Nonlinearity versus frequency is virtually impossible to measure directly, but instruments such as distortion analyzers make it relatively easy to measure harmonic distortion versus frequency. Like feedthrough, multiplier distortion usually increases with frequency, and the amount of distortion may be very different for the x and y inputs.

For very low frequencies, the percent harmonic distortion is usually equal to the dc nonlinearity as a percent of 10 V divided by $\sqrt{2}$, i.e.,

$$\text{Percent distortion} \doteq \frac{\text{percent nonlinearity}}{\sqrt{2}}$$

This approximation is valid for most multipliers for frequencies well below the -3 dB small-signal frequency.

SPECIFICATIONS

Typical performance at +25°C with rated power supplies unless otherwise noted.
Per cent specifications refer to % of full scale (10V).

ELECTRICAL

OUTPUT FUNCTION	XY/10
TOTAL ERROR	
Internal trim	1%, max.
External trim	0.6%
vs. Temperature	0.04%/°C
vs. Supply	0.2%/%
INDIVIDUAL ERRORS	
Output Offset @ 25°C (X=Y=0)	20 mV, max.
vs. Temperature (Operating Range)	0.4 mV/°C
vs. Supply	10 mV/%
Scale Factor Error	0.6%
vs. Temperature (Operating Range)	0.04%/°C
vs. Supply	0.1%/%
Non-Linearity	
X (X=20 V p-p, Y=±10VDC)	0.5%
Y (Y=20 V p-p, X=±10VDC)	0.2%
Feedthrough @ 50 Hz	
X=0, Y=20V p-p (Internal Trim)	50 mV p-p
(External Trim)	20 mV p-p
vs. Temperature	1 mV p-p/°C
Y=0, X=20V p-p (Internal Trim)	50 mV p-p
(External Trim)	20 mV p-p
vs. Temperature	2 mV p-p/°C
AC PERFORMANCE	
Slew Rate	25 V/μsec
-3 dB Small Signal Bandwidth	1 MHz
1% Amplitude Error	40 kHz
1% Vector Error (0.57° phase shift)	10 kHz
Settling Time (2% of final value, 20V, step)	1 μsec
Overload Recovery Time	3 μsec
OUTPUT NOISE (X=Y=0)	
10 kHz to 10 MHz	3 mV rms
10 Hz to 10 kHz	600 μV rms
INPUT CHARACTERISTICS	
Input Voltage Range	
Rated Operation	±10 V
Absolute Max.	±15 V
Input Impedance, X	10 MΩ
Y	10 MΩ
Z	36 KΩ
OUTPUT CHARACTERISTICS	
Rated Output	±10 V @ ±5 mA
Output Impedance	1 Ω
POWER SUPPLY REQUIREMENTS	
Rated Voltage	±15 VDC
Operating Range	±12 VDC to ±18 VDC
Quiescent Current	±4 mA
TEMPERATURE RANGE	
Operating, Rated Performance	-55° to +125°C
Storage	-65°C to +150°C

Fig. 3.5 A typical multiplier specification sheet.

MECHANICAL
TO-100

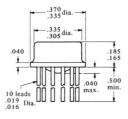
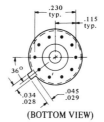

(BOTTOM VIEW)

TYPICAL PERFORMANCE CURVES
Typical Performance @ 25°C and ±15 VDC

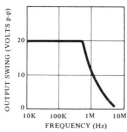

FIGURE 1. Large Signal Frequency Response.

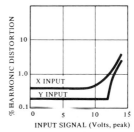

FIGURE 4. Output Distortion vs. Peak Input Signal.

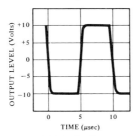

FIGURE 2. Step Response

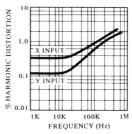

FIGURE 5. Output Distortion vs. Frequency.

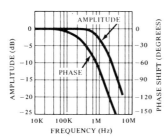

FIGURE 3. Small Signal Frequency Response.

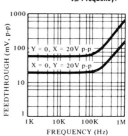

FIGURE 6. AC Feedthrough vs. Frequency.

82 Function Circuits

Differential phase shift. The phase shifts for x input signals and y input signals are rarely equal. The difference between ϕ_x and ϕ_y is called *differential phase shift* ϕ_D. In modulation applications ϕ_D can generate a dc output error term. However, unless ϕ_D is several degrees, the dc error will generally be insignificant compared to the multiplier dc output offset voltage. Unless the multiplier has significantly different bandwidths for the x and y inputs, ϕ_D is usually an unimportant error; in most cases it is not specified.

3.1.4 A typical multiplier specification sheet We have discussed the definitions of various multiplier specifications. A typical specification sheet is given in Fig. 3.5. It is an integrated-circuit transconductance multiplier designed for low-cost general-purpose usage. The multiplier is laser trimmed prior to final packaging and is specified at 1 percent maximum total error with no external components. Optional adjustments may be made to reduce the total error further, typically down to 0.6 percent.

The single-chip integrated-circuit design makes it possible for the multiplier to be packaged in a hermetically sealed TO-100 can, capable of being operated in adverse environmental conditions and over a wide temperature range.

3.2 Test Multipliers Using Crossplot Technique

To make accurate multiplier adjustments and to measure multiplier error terms, especially nonlinearity, the crossplot technique is most useful. It presents a concise error curve on an oscilloscope or a $x - y$ plotter. The circuit diagram in Fig. 3.6 illustrates the general setup to measure total error, gain error, and nonlinearity. The two R_2 input resistors in the summing amplifier must be matched to within 0.02 percent, as must the two R_1 resistors in the inverting amplifier.

Typically a ± 10 V, 5 Hz sine wave is connected to one input of the multiplier and also to the horizontal input of the oscilloscope. If the multiplier under test has a narrow bandwidth, it may be necessary to lower the sine-wave generator frequency; otherwise a hysteresis loop would show up in the error curve because of the phase shift through the multiplier. The other multiplier input is connected to a precision dc reference voltage, $+10$ or -10 V. When the dc reference voltage is at -10 V, the inverting amplifier should be bypassed, and so the switch S should be turned to position 2. The output of the summing amplifier (sometimes called the error amplifier) is the instantaneous error of the multiplier. On the oscilloscope, gain error will show up as a rotation of the error trace, and any nonlinearity will show up as bends in the error trace. When the dc reference is at $+10$ V, the switch S must be turned to position 1. This

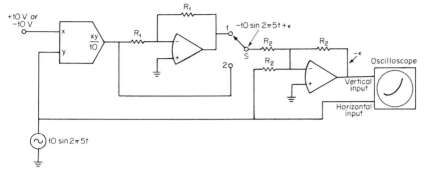

Fig. 3.6 Crossplot technique to measure multiplier total error, gain error, and nonlinearity.

method of measurement depends upon the accuracy of the dc reference voltages and upon the errors from the inverting and error amplifiers but is not dependent upon the amplitude of the sine-wave generator. Some typical error curves for a general-purpose variable transconductance multiplier are given in Fig. 3.7.

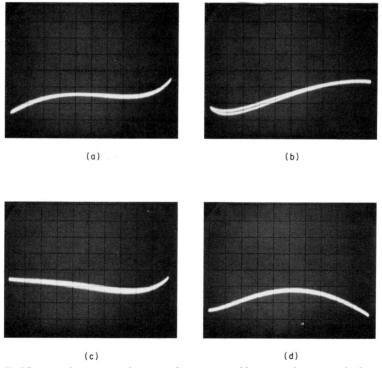

Fig. 3.7 Typical error curves for a general-purpose variable transconductance multiplier; vertical scale = 20 mV/division, horizontal scale = 2 V/division. (a) $x = +10$ V dc, $y = 10 \sin 2\pi 5t$; (b) $x = -10$ V dc, $y = 10 \sin 2\pi 5t$; (c) $x = 10 \sin 2\pi 5t$, $y = +10$ V dc; (d) $x = 10 \sin 2\pi 5t$, $y = -10$ V dc.

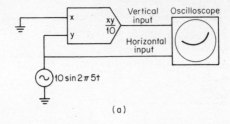

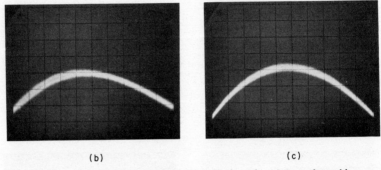

Fig. 3.8 (a) Setup for multiplier feedthrough crossplot. (b), (c) Typical variable transconductance multiplier feedthrough with (b) $x = 0$, $y = 10 \sin 2\pi 5t$, and (c) $y = 0$, $x = 10 \sin 2\pi 5t$. Vertical scale = 20 mV/division; horizontal scale = 2 V/division.

The setup to crossplot feedthrough is much simpler (see Fig. 3.8a). One input of the multiplier is grounded, while the other input is connected to a ±10 V sine-wave generator. By adjusting the respective input offset (x input or y input), the feedthrough trace can be rotated until the two end points come to the same horizontal level. The remaining bend in the trace is due to the nonlinearity effect on the feedthrough and can be trimmed out only on certain types of multipliers. Typical feedthrough curves for a variable transconductance multiplier are given in Fig. 3.8b and c.

3.3 Multiplication Techniques

The best known and most popular of today's packaged analog circuits, excluding operational amplifiers, is the analog multiplier. Several all-electronic techniques of multiplication are currently in use, including log-antilog, variable-transconductance, pulse width/pulse height, triangle-averaging, and quarter-square. Sophisticated hybrid and monolithic technologies have already penetrated the analog multiplier industry, resulting in considerable cost and performance improvement. We will first discuss the general fundamentals of various multiplication techniques and then go into the details of the circuit theories. A complete adjustment procedure is described for the log-antilog multiplier. The same procedures can be modified to apply to other types of multipliers.

3.3.1 General fundamentals and a comparison chart

The first commercial analog multiplier in modular form utilized the quarter-square technique. This technique is based on the algebraic identity

$$\frac{1}{4}[(x+y)^2 - (x-y)^2] = xy$$

Only two types of elements are involved: summing amplifiers and squaring circuits. The precision squaring circuits available at the time such multipliers were introduced were based on the piecewise-linear approximation technique. This technique uses diode breakpoint networks. The resulting error appears as a ripple function of input voltage. The most serious drawback of the quarter-square multiplier is that it has relatively large errors for small inputs, even though the maximum error is a small percentage of full scale. The major advantage is its excellent frequency response.

Two multiplying techniques closely related in their performance characteristics are the triangle-averaging method and the pulse width/pulse height modulation technique. Both require a triangle-waveform generator for the carrier, operating at a frequency well above the frequency response of the multiplier, and a built-in low-pass filter to remove this carrier frequency from the output. The triangle-averaging technique uses precision rectifiers as demodulators, while the pulse width/pulse height method uses switches. The triangle waveform is required to be symmetrical, linear, and sharp-peaked to achieve accurate multiplication by either technique. Given such waveforms, however, these multipliers are capable of accuracies to within 0.1 percent or better. Frequency response is somewhat limited because of carrier frequency and switching-time limitations and the necessity for low-pass filtering at the output. Bandwidths above 1 kHz are nevertheless quite common and are sufficient for many applications.

The first multipliers realized in integrated-circuit (IC) form made use of the inherent close matching of transistors fabricated on a single chip and the variable transconductance of the semiconductor junction. This is the so-called *variable-transconductance multiplier*. Accurate multiplication requires that the transistors be dynamically matched. This is difficult, but it can be done by selecting pairs of discrete transistors. It is much easier, however, to match transistors and resistors by design—that is, to build them with identical geometries and in close proximity on a single chip of silicon. This ensures excellent tracking of parameters over time, temperature, and supply variations and, at the same time, yields low cost and

TABLE 3.1 Comparison Chart for Different Types of Multipliers

Multiplication technique	Quarter-square	Pulse width/ pulse height	Triangle-averaging	Variable-transconductance	Log-antilog
Total error, % of full scale (10 V)	0.25–0.5%	0.01–0.1%	0.1–0.5%	0.5–2%	0.1–0.5%
Total error vs. temperature	0.03%/°C	0.01%/°C	0.03%/°C	0.02%/°C	0.015%/°C
−3 dB small-signal bandwidth	2 MHz	200 Hz	1 kHz	2 MHz	250 kHz
Slew rate	3 V/μs	0.7 V/ms	5 V/ms	3 V/μs	1 V/μs
General comments	High accuracy, wide bandwidth	Excellent static accuracy, low bandwidth	High accuracy, slow response; made obsolete by log-antilog	Moderate accuracy, wide bandwidth, general-purpose, low cost	High accuracy, medium bandwidth, low cost

reliability. The first such IC multipliers had serious limitations on their ease of use, supply sensitivity, and accuracy. Later versions removed some of these limitations, but still needed considerable trimming and adjustments. Recently, however, IC multipliers using thin-film technology and laser trimming have eliminated all external adjustments and have performances fully comparable with those of many modular multipliers.

The IC variable-transconductance multiplier has advantages over the previously mentioned techniques, namely, good transient response, smooth error curves (i.e., low error for small signals), and low noise. At present it is limited to about 0.5 percent error. However, with improved processing and device matching, and by combining monolithic, thin-film and laser trimming technologies, it seems fairly certain that performance improvements will be forthcoming in the near future.

The log-antilog technique uses the property of semiconductor junctions in a somewhat different way. This technique utilizes the logarithmic voltage-versus-current relationship in semiconductor junctions. Transistor base-emitter junctions have been found to yield excellent log conformity, and antilogging can be performed by simply changing the interconnections of the logging transistor. By combining logging and antilogging transistors, operational amplifiers, and suitable resistor networks, one can perform multiplication and division. If thin-film resistor technology is used and all four logging and antilogging transistors are designed on the same monolithic silicon chip for parameter matching and tracking, this approach yields high accuracy, relatively wide bandwidth, and low cost. Recently, the log-antilog multipliers have become commercially available in a miniature 14-pin hybrid package with maximum total error of 0.25 percent. This technique has yet another feature: the capability of performing accurate division that cannot be achieved at the present time by other techniques.

A comparison chart is given in Table 3.1 to summarize the typical performance obtainable with different multiplication techniques. The values are listed as a measure of relative performance. They may be somewhat loose in some places and tight in others, depending on the specific design. For example, if bandwidth is of primary concern, one can design a variable transconductance multiplier with 10 MHz small-signal 3 dB bandwidth at the expense of static accuracy and cost.

3.3.2 Quarter-square multiplier

Quarter-square multiplication is based upon the algebraic identity

$$xy = \frac{1}{4}[(x+y)^2 - (x-y)^2]$$

One-quarter of the difference of the two squares equals the product xy; hence the name *quarter-square multiplier* is obtained. A scale factor of 1:10 is used to make the output voltage range equal to the input voltage range of ± 10 V. Thus

$$e_o = \frac{xy}{10}$$

$$= \frac{1}{40}[(x+y)^2 - (x-y)^2]$$

Diode shaping networks are used to perform the squaring operation. Three operational amplifiers and two squaring circuits are used in each multiplier. The circuit diagram is given in Fig. 3.9a, with the positive

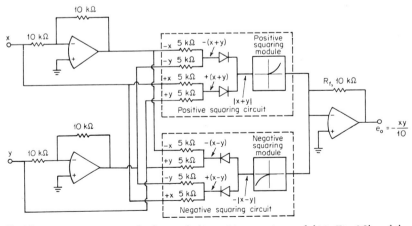

Fig. 3.9a. A quarter-square multiplier, with the positive squaring module in Fig. 3.9b and the negative squaring module in Fig. 3.9c.

88 Function Circuits

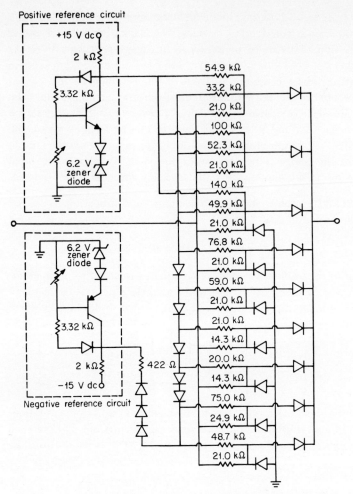

Fig. 3.9b. The positive squaring module for the quarter-square multiplier shown in Fig. 3.9a (negative squaring module shown in Fig. 3.9c).

squaring module in Fig. 3.9b and the negative squaring module in Fig. 3.9c. Note that there is one positive reference circuit and one negative reference circuit in each squaring module. In practice, only one positive and one negative reference circuit are needed to supply both the positive and negative squaring modules.

Each squaring circuit receives four inputs: x, y, $-x$, and $-y$. x and y are obtained externally, while $-x$ and $-y$ are obtained from the internal inverting amplifiers. In the positive squaring circuit, signals proportional to $(x + y)$ and $-(x + y)$ are formed with precision resistors. The more positive signal is selected by a pair of rectifying diodes and used to drive the positive squaring module, thus providing an output current proportional to $(x + y)^2$. Similarly, in the negative squaring circuit $(x - y)$ and $-(x - y)$ are formed, and the more negative signal is used to drive the negative squaring module that provides an output current proportional to $-(x - y)^2$.

When these two currents, $(x + y)^2$ and $-(x - y)^2$, are summed at the summing junction of the output operational amplifier, the result is a current proportional to $4xy$, as

$$(x + y)^2 - (x - y)^2 = 4xy$$

Each term on the left contributes its error to the sum; thus the squaring circuits must be twice as accurate as the multiplier. The squaring circuit

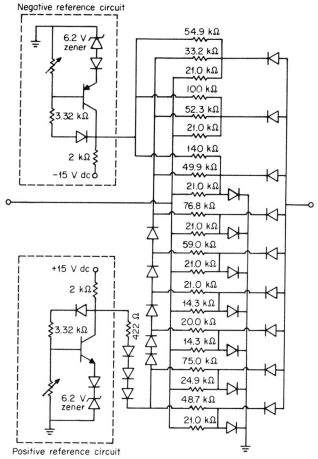

Fig. 3.9c. The negative squaring module for the quarter-square multiplier shown in Fig. 3.9a (positive squaring module shown in Fig. 3.9b).

90   Function Circuits

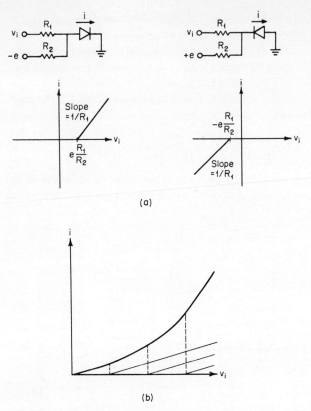

Fig. 3.10 (a) Ideal limiter channel input-output characteristics. (b) Composite diode limiter channel characteristics.

transfer characteristics yield an output voltage equal to $-xy/10$ when $R_f = 10$ kΩ.

Diode squaring modules. The positive and negative squaring modules are made of diode limiters. Each module contains a diode function generator composed of several diode limiter channels in parallel. Simplified diagrams of two typical limiter channels are shown in Fig. 3.10a. If several diode limiter channels are connected in parallel and their inputs are all driven by the same voltage and their output currents are all summed at a common node, then a composite transfer characteristic may be drawn, as shown in Fig. 3.10b.

3.3.3 Pulse width/pulse height multiplier Pulse time modulation is one of the most accurate analog multiplication techniques. However, it has the drawback of slow frequency response, restricting it to static applications only. The theory of operation of pulse width/pulse height multipliers is based on the principle that one input voltage, say v_x, controls the width or duty cycle of a pulse train, while the other input voltage, v_y, is made proportional to the amplitude of the pulse train. It then passes through a low-pass filter which retains only the dc component of the pulse train. The output dc voltage is proportional to the product of v_x and v_y.

Pulse time multipliers can be grouped into two categories: externally excited, which require an external oscillator, and internally excited, which oscillate of their own accord. We will discuss a special kind of externally excited pulse time multiplier which provides accuracies of ± 0.01 percent at $\pm 25°C$ and ± 0.1 percent from -55 to $+125°C$.

In addition to the external triangle-wave generator, the pulse time multiplier consists mainly of three blocks: a pulse width modulator, a pulse height modulator, and a low-pass filter, as shown in Fig. 3.11a. The various output waveforms and their relationship to one another are plotted

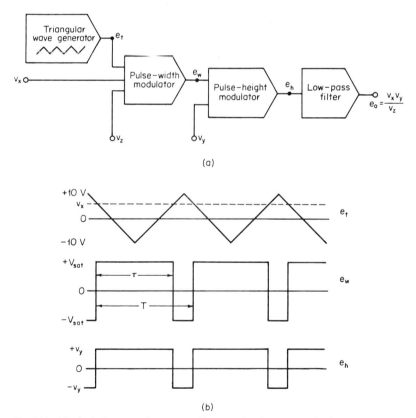

Fig. 3.11 (a) Block diagram of an externally excited pulse time multiplier. (b) Output waveforms of the triangle-wave generator e_t, the pulse width modulator e_w, and the pulse height modulator e_h.

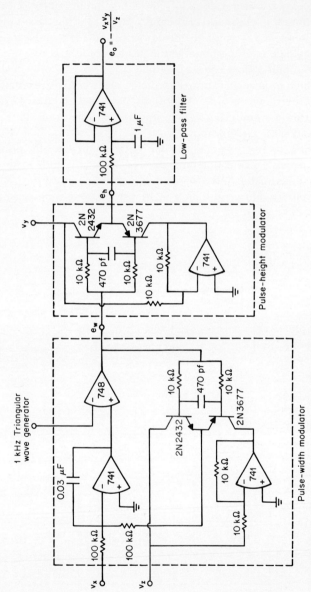

Fig. 3.12 Precision four-quadrant pulse width pulse height multiplier.

in Fig. 3.11b. The pulse width modulator can be considered as a four-terminal device with three inputs e_t, v_x, v_z and an output e_w. v_z is a reference voltage, usually $+10$ V. The modulator's output e_w saturates to $+V_{sat}$ ($+12$ V if a 748-type amplifier is used as its comparator) when $e_t < v_x$ and saturates to $-V_{sat}$ when $e_t > v_x$. From the first two plots in Fig. 3.11b, it can be verified graphically that

$$\frac{v_x}{v_z} = \frac{2\tau}{T} - 1$$

where $v_z = +10$ V, τ is the duty cycle of the pulse train e_w, and T is the period. Then e_w goes into the pulse height modulator, whose output pulse train e_h is identical to e_w except for being amplitude-limited by $\pm v_y$. The dc component of e_h which appears at the output of the low-pass filter is

$$e_o = v_y\left(\frac{2\tau}{T} - 1\right) \tag{3-4}$$

Substituting Eq. (3-3) into (3-4), we have

$$e_o = \frac{v_x v_y}{v_z} \tag{3-5}$$

The complete circuit diagram is given in Fig. 3.12. In order to achieve high accuracies, the triangle-wave-generator frequency should be kept as low as the specific application allows, preferably below 1 kHz. The pulse width modulator is made up of an input integrator, a comparator, an electronic switch, and an inverting amplifier.

Identical bipolar analog switches are used in both the pulse width and pulse height modulators. Each switch consists of two silicon transistors, 2N2432 and 2N3677, two 10 kΩ resistors, and one 470 pF capacitor. The bipolar transistors must have breakdown voltages (BV_{CBO}, BV_{EBO}, BV_{CEO}) of over 25 V, leakage currents (I_{CBO}, I_{EBO}) of less than 1 μA, dynamic saturation resistance of less than 20 Ω, offset voltage of less than ±2 mV, and switching times (t_r, t_d, t_{st}, t_f) of less than 200 ns.

A 748-type operational amplifier is called out as the comparator to provide the ±12 V drive required by the analog switches. Its input is over-voltage-protected. Standard 741-type amplifiers are used elsewhere for low cost. The amplifiers' offsets should be zero-adjusted for better accuracies.

3.3.4 Triangle-averaging multiplier

The block diagram of a triangle-averaging multiplier is given in Fig. 3.13a. It is made up of a triangle-wave generator, summing amplifiers, diode rectifiers, and low-pass filters.

94 Function Circuits

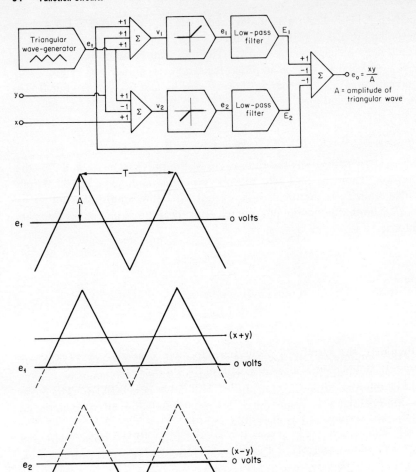

Fig. 3.13 (a) Block diagram of a triangle-averaging multiplier. (b) Output waveforms of the triangle-wave generator e_t, of the positive diode rectifier e_1, and of the negative diode rectifier e_2.

The triangle wave e_t and inputs x and y are first combined through two separate summing amplifiers in such a way that their outputs are

$$v_1 = e_t + x + y$$

and

$$v_2 = e_t + x - y$$

Then v_1 passes into a diode rectifier which only retains its positive portion. The waveform of the rectifier's output, e_1, is drawn in heavy lines in Fig.

3.13b. The following low-pass filter extracts the dc component of e_1 such that its output E_1 is

$$E_1 = \frac{1}{T}\int_0^t e_1(t)\, dt$$

$$= \frac{1}{4A}(x + y + A)^2$$

Similarly, the output E_2 of the other channel is

$$E_2 = \frac{1}{4A}(-x + y + A)^2$$

E_1, E_2, and x are combined through a summing amplifier in such a way that the output e_o becomes

$$e_o = E_1 - E_2 - x$$

Hence

$$e_o = \frac{xy}{A}$$

Obviously, the amplitude of the triangle-wave generator, A, must be extremely stable as it controls the gain of the multiplier.

The circuit diagram of a complete triangle-averaging multiplier is given in Fig. 3.14b, with the 12 kHz triangle-wave generator shown in Fig. 3.14a and the 1 kHz low-pass filter in Fig. 3.14c. Note that only one low-pass filter is needed if it is connected after the summing amplifier. The circuit's static accuracy is typically ±0.25 percent of full scale, and the

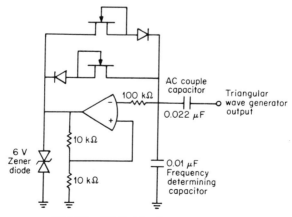

Fig. 3.14a. 12 kHz triangle-wave generator.

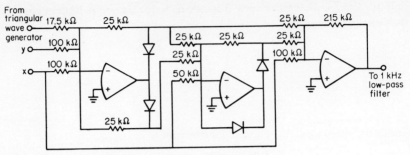

Fig. 3.14b. Circuit diagram of a triangle-averaging multiplier.

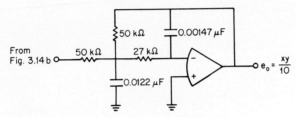

Fig. 3.14c. Low-pass filter with -3 dB frequency at 1 kHz, Q factor $= 1$.

drift is $\pm 0.03\%/°C$. The frequency response is limited, with ± 1 percent amplitude error at about 100 Hz.

3.3.5 Variable-transconductance multiplier

The variable transconductance of silicon junction semiconductors is widely utilized in performing analog multiplications. This class of multipliers provides wide bandwidth, differential inputs, good linearity, and low cost. Because the circuitry is relatively simple compared with that for other multiplication techniques and because the accuracy depends on a close match of adjacent transistors, integrated-circuit construction is especially desirable. The majority of IC multipliers in the commercial market use this technique.

A four-quadrant variable-transconductance multiplier basically consists of a multiplier core, x and y emitter-degenerated amplifiers, and an output difference amplifier. Figure 3.15 shows a simplified circuit diagram. Three pairs of silicon junction transistors, Q_{1A}/Q_{1B}, Q_{2A}/Q_{2B}, and Q_{3A}/Q_{3B}, make up the multiplier core. The x and y input amplifiers are used to convert input voltages into pairs of complementary currents. Each amplifier requires a pair of closely matched transistors and a highly stable resistor. Q_{4A}/Q_{4B} and R_x are for the x amplifier, and Q_{5A}/Q_{5B} and R_y are for the y amplifier. The output of the difference operational amplifier is proportional to the product of $(x_1 - x_2)$ and $(y_1 - y_2)$.

We will now derive a transfer function between the output and the inputs for the multiplier circuit. The logarithmic equation (see Chapter 2 for details) of a bipolar transistor is

$$v_{BE} = \frac{kT}{q} \ln \frac{i_c}{I_S} \tag{3-6}$$

where v_{BE} = base-emitter voltage
k = Boltzmann's constant
q = charge of electron
T = absolute temperature
i_c = collector current
I_S = saturation current

Apply Eq. (3-6) to transistors Q_{1A} and Q_{1B}; then

$$V_1 - V_2 = \frac{kT_{1A}}{q} \ln \frac{i_{1A}}{I_{S1A}} \tag{3-7}$$

and

$$V_1 - V_3 = \frac{kT_{1B}}{q} \ln \frac{i_{1B}}{I_{S1B}} \tag{3-8}$$

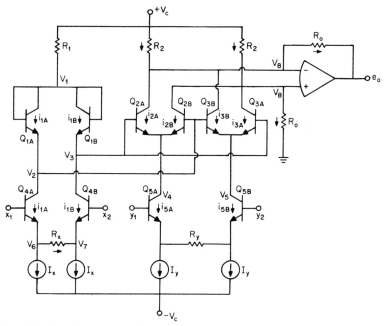

Fig. 3.15 Simplified circuit diagram of variable-transconductance multiplier, $e_o = \dfrac{2R_o}{I_x R_x R_y}(x_1 - x_2)(y_1 - y_2)$.

Function Circuits

Assume Q_{1A} and Q_{1B} are a matched pair; then $T_{1A} = T_{1B} = T_1$ and $I_{S1A} = I_{S1B} = I_{S1}$. Subtracting Eq. (3–8) from (3–7) yields

$$V_3 - V_2 = \frac{kT_1}{q} \ln \frac{i_{1A}}{i_{1B}} \tag{3-9}$$

Similarly, applying Eq. (3–6) to transistor pairs Q_{2A}/Q_{2B} and Q_{3A}/Q_{3B}, respectively, and using the same simple algebra, the following two equations result:

$$V_3 - V_2 = \frac{kT_2}{q} \ln \frac{i_{2A}}{i_{2B}} \tag{3-10}$$

and

$$V_3 - V_2 = \frac{kT_3}{q} \ln \frac{i_{3A}}{i_{3B}} \tag{3-11}$$

Further assume that the three pairs Q_{1A}/Q_{1B}, Q_{2A}/Q_{2B}, and Q_{3A}/Q_{3B} are adjacent to one another and that they have the same component temperature; then $T_1 = T_2 = T_3 = T$. Equating Eqs. (3–9), (3–10), and (3–11) yields

$$\frac{i_{1A}}{i_{1B}} = \frac{i_{2A}}{i_{2B}} = \frac{i_{3A}}{i_{3B}} \tag{3-12}$$

Assuming an ideal output operational amplifier, the voltage at the inverting input, V_8, is the same as the voltage at the noninverting input. Sum up all the currents at the inverting input:

$$\frac{V_c - V_8}{R_2} = i_{2A} + i_{3B} + \frac{V_8 - e_o}{R_o} \tag{3-13}$$

Similarly, at the noninverting input,

$$\frac{V_c - V_8}{R_2} = i_{2B} + i_{3A} + \frac{V_8}{R_o} \tag{3-14}$$

From Eqs. (3–13) and (3–14), we have

$$\frac{e_o}{R_o} = i_{2A} - i_{2B} - i_{3A} + i_{3B} \tag{3-15}$$

Assuming Q_{4A}/Q_{4B} is a matched pair and applying Eq. (3–6) to Q_{4A} and Q_{4B}, respectively, we obtain

$$x_1 - V_6 = \frac{kT_4}{q} \ln \frac{i_{1A}}{I_{S4}} \tag{3-16}$$

and

$$x_2 - V_7 = \frac{kT_4}{q} \ln \frac{i_{1B}}{I_{S4}} \qquad (3\text{-}17)$$

Subtract Eq. (3-17) from (3-16) and simplify:

$$x_1 - x_2 = V_6 - V_7 + \frac{kT_4}{q} \ln \frac{i_{1A}}{i_{1B}} \qquad (3\text{-}18)$$

Sum up all currents at the emitters of Q_{4A} and Q_{4B}, respectively:

$$i_{1A} = \frac{V_6 - V_7}{R_x} + I_x \qquad (3\text{-}19)$$

and

$$i_{1B} + \frac{V_6 - V_7}{R_x} = I_x \qquad (3\text{-}20)$$

Add Eqs. (3-19) and (3-20) together:

$$i_{1A} + i_{1B} = 2I_x \qquad (3\text{-}21)$$

Substitute $(V_6 - V_7)$ from Eq. (3-19) into (3-18):

$$x_1 - x_2 = (i_{1A} - I_x)R_x + \frac{kT_4}{q} \ln \frac{i_{1A}}{i_{1B}} \qquad (3\text{-}22)$$

From Eqs. (3-21) and (3-22),

$$x_1 - x_2 = \frac{R_x}{2}(i_{1A} - i_{1B}) + \frac{kT_4}{q} \ln \frac{i_{1A}}{i_{1B}} \qquad (3\text{-}23)$$

To simplify the derivation, neglect the logarithmic term in Eq. (3-23). It contributes only a small amount of nonlinearity error. Equation (3-23) then becomes

$$x_1 - x_2 = \frac{R_x}{2}(i_{1A} - i_{1B}) \qquad (3\text{-}24)$$

A similar derivation can be made on the y input amplifier such that

$$y_1 - y_2 = \frac{R_y}{2}(i_{5A} - i_{5B}) \qquad (3\text{-}25)$$

From Eqs. (3-24) and (3-25),

$$\frac{4(x_1 - x_2)(y_1 - y_2)}{R_x R_y} = (i_{1A} - i_{1B})(i_{5A} - i_{5B}) \qquad (3\text{-}26)$$

Function Circuits

Sum up the currents at the collectors of Q_{5A} and Q_{5B}, respectively:

$$i_{5A} = i_{2A} + i_{2B} \tag{3-27}$$

and

$$i_{5B} = i_{3A} + i_{3B} \tag{3-28}$$

From Eqs. (3-21) and (3-12), i_{1A} and i_{1B} can be formulated in terms of i_{2A}, i_{2B}, and I_x:

$$i_{1A} = 2I_x \frac{i_{2A}}{i_{2A} + i_{2B}} \tag{3-29}$$

and

$$i_{1B} = 2I_x \frac{i_{2B}}{i_{2A} + i_{2B}} \tag{3-30}$$

Substitute Eqs. (3-27), (3-28), (3-29), and (3-30) into (3-26) and simplify:

$$\frac{4(x_1 - x_2)(y_1 - y_2)}{R_x R_y} = 2I_x \left[i_{2A} - i_{2B} - (i_{3A} + i_{3B}) \frac{i_{2A} - i_{2B}}{i_{2A} + i_{2B}} \right] \tag{3-31}$$

From Eq. (3-12), it can be verified that

$$\frac{i_{2A} - i_{2B}}{i_{2A} + i_{2B}} = \frac{i_{3A} - i_{3B}}{i_{3A} + i_{3B}} \tag{3-32}$$

Substitute Eq. (3-32) into (3-31) and simplify:

$$\frac{4(x_1 - x_2)(y_1 - y_2)}{R_x R_y} = 2I_x (i_{2A} - i_{2B} - i_{3A} + i_{3B}) \tag{3-33}$$

Comparing Eq. (3-33) with (3-15), it is obvious that

$$a_0 = \frac{2R_v}{I_x R_x R_y}(x_1 - x_2)(y_1 - y_2) \tag{3-34}$$

The circuit diagram of a complete variable-transconductance multiplier is given in Fig. 3.16. Total error of 1 percent of full scale (10 V) is easily achievable.

If the feedback around the output operational amplifier is through an extra z amplifier identical to the x and y input amplifiers, as in Fig. 3.17, the nonlinearity error due to the logarithmic term in Eq. (3-23) and also that of the y input amplifier will be cancelled out. This approach also has several other advantages:

1. The sensitivity of output offset voltage to supply voltages is very low.
2. As a feedback divider, both numerator and denominator are fully differential and have high input impedances.

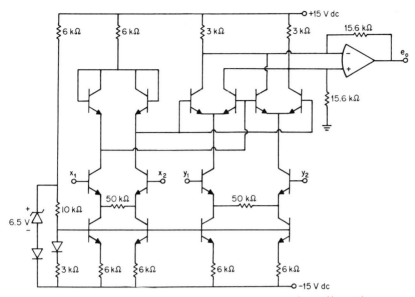

Fig. 3.16 Four-quadrant variable-transconductance multiplier, $e_o = \dfrac{(x_1 - x_2)(y_1 - y_2)}{10}$

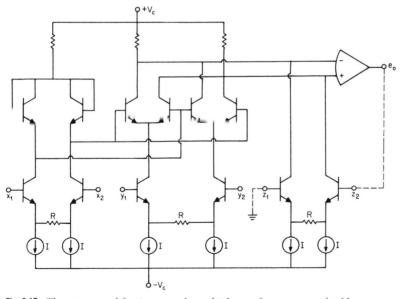

Fig. 3.17 The extra z amplifier improves the multiplier performance considerably.

101

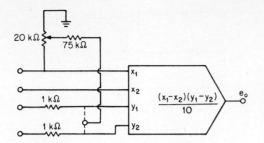

Fig. 3.18 External linearity compensation on variable-transconductance multipliers.

3. The unused feedback terminal permits the addition of a signal to the output.
4. The transient response is improved because of the cancellation of equal zeros in the emitter circuits of the input and feedback stages.
5. The scale factor can be readily changed by an additional feedback attenuator.

Linearity compensation to improve accuracy. External linearity compensation can be made on commercial variable-transconductance multipliers to reduce errors. Total error, nonlinearity, and feedthrough can be improved up to an order of magnitude. The compensation is based on the theory that if a fraction of the x input is cross-coupled into the y input signal, the second-order errors caused by offset in the multiplier core will be cancelled out. The y input can also be linearized by cross-coupling to x, but this is normally not required because the y input nonlinearity is so low that nonreducible errors, caused by third-order terms, are as high as second-order errors.

Figure 3.18 illustrates the compensation connections. Use the procedures discussed in Sec. 3.2 to crossplot the nonlinearity error curve of feedthrough. Connect the 75 kΩ resistor to either the y_1 input or the y_2 input, depending on whether the error trace curves downward or upward. This compensation scheme does not affect other specifications of the multiplier such as temperature drift or frequency response. The x feedthrough is reduced at direct current as well as at high frequencies.

3.3.6 Log-antilog multiplier Recent developments have made it possible for log-antilog multipliers to yield excellent accuracy, approaching that of pulse height/pulse width multipliers, but with much wider bandwidth, low noise, and reduced cost. Accuracy of 0.1 percent is achievable. A log-antilog multiplier mainly consists of four operational amplifiers and two pairs of silicon junction transistors to perform the logging and antilogging functions. A simplified one-quadrant log-antilog multiplier circuit diagram is given in Fig. 3.19. Rewrite Eq. (3-6), the logarithmic equation for bipolar transistors, as

$$v_{BE} = \frac{kT}{q} \ln \frac{i_c}{I_S} \tag{3-35}$$

Apply Eq. (3-35) to logging transistors Q_{1A} and Q_{1B}, respectively:

$$0 - V_1 = \frac{kT_{1A}}{q} \ln \frac{x/R_x}{I_{S1A}} \tag{3-36}$$

and

$$0 - V_3 = \frac{kT_{1B}}{q} \ln \frac{z/R_z}{I_{S1B}} \tag{3-37}$$

Assume Q_{1A}/Q_{1B} is a matched pair, so that $T_{1A} = T_{1B} = T_1$ and $I_{S1A} = I_{S1B} = I_{S1}$. Subtract Eq. (3-37) from (3-36) and simplify.

$$-V_1 + V_3 = \frac{kT_1}{q} \ln \frac{xR_z}{zR_x} \tag{3-38}$$

Similarly, assume Q_{2A}/Q_{2B} is a matched pair and apply Eq. (3-35) to Q_{2A} and Q_{2B}, respectively:

$$V_1 - V_2 = \frac{kT_2}{q} \ln \frac{y/R_y}{I_{S2}} \tag{3-39}$$

and

$$V_3 - V_2 = \frac{kT_2}{q} \ln \frac{e_o/R_o}{I_{S2}} \tag{3-40}$$

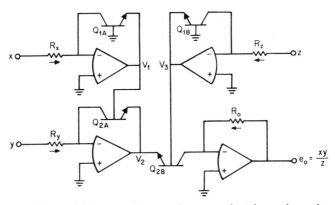

Fig. 3.19 Simplified circuit diagram of a one-quadrant log-antilog multiplier.

Subtracting Eq. (3–39) from (3–40) and simplifying results yields

$$-V_1 + V_3 = \frac{kT_2}{q} \ln \frac{e_o R_y}{y R_o} \quad (3\text{–}41)$$

Further assume that the two pairs, Q_{1A}/Q_{1B} and Q_{2A}/Q_{2B}, are placed adjacent to each other such that $T_1 = T_2 = T$. Then, from Eqs. (3–38) and (3–41), the following equation results:

$$e_o = \frac{xy}{z} \frac{R_z R_o}{R_x R_y}$$

Make $R_z = R_o = R_x = R_y$; then

$$e_o = \frac{xy}{z} \quad (3\text{–}42)$$

To use the circuit as a multiplier, z is connected to a +10 V dc stable reference voltage. As a divider, y is connected to a +10 V dc stable reference voltage. The x, y, and z inputs must be positive voltages, i.e., one-quadrant operation only is possible. The circuit diagram for a complete high-performance four-quadrant log-antilog multiplier is given in Fig. 3.20. R_2, R_4, R_{14}, R_{15}, and R_{16} are level-shifting resistors for conversion from one-quadrant to four-quadrant operation. Since a stable +10 V reference source requires an additional operational amplifier, the 6.2 V temperature-compensated zener diode DZ is used to supply the z input. To match the 6.2 V reference, R_2, R_4, R_{16}, and R_{18} are made 62 kΩ. C_1/R_6, C_2/R_7, and C_3/R_{19} are phase compensation networks to stabilize amplifiers A_1, A_2, and A_3, respectively. Diodes D_1, D_2, D_3, and D_4 are used to protect the emitter-base junctions of Q_{1A}, Q_{1B}, Q_{2A}, and Q_{2B}, respectively, in case of accidental reverse biasing. Q_{1A}/Q_{1B} and Q_{2A}/Q_{2B} must be matched pairs, and all four transistors must be at the same temperature. R_{19}, R_{20}, R_{10}/R_{11}, R_{12}/R_{13}, and R_8/R_9 are linearity compensation resistors to trim out x feedthrough nonlinearity, y feedthrough nonlinearity, and gain nonlinearity. These eight resistors are not required if only a 1 percent multiplier is needed, but to achieve 0.1 percent accuracy, linearity compensations must be performed.

Adjustment procedures:

1. Output offset: Ground x and y inputs and adjust R_4 and/or R_{18} for $e_o = 0$.
2. (a) y input offset or y feedthrough: Set up the equipment as in Fig. 3.8a, Sec. 3.2, to crossplot the feedthrough on an oscilloscope. Ground x and connect y to the ±10 V, 5 Hz sine-wave generator. Adjust R_5 and/or R_{15} such that the two end points of the oscilloscope trace are in the same horizontal level.

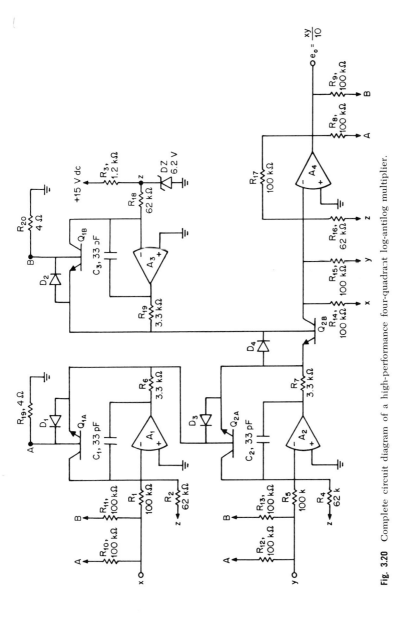

Fig. 3.20 Complete circuit diagram of a high-performance four-quadrant log-antilog multiplier.

106 Function Circuits

   (b) y feedthrough linearity compensation: Adjust R_{12} and/or R_{13} such that the U-shaped oscilloscope trace turns into a rotated sine-shaped trace.

   (c) Repeat step 2a until the rotated sine-shaped trace turns to a regular horizontal sine-shaped trace. The remaining nonlinearity is due to third-order terms and cannot be further reduced.

3. (a) x input offset or x feedthrough: The equipment setup is similar to that in step 2 to crossplot the x feedthrough. Ground y and connect x to the ± 10 V, 5 Hz sine-wave generator. Adjust R_1 and/or R_{14} such that the two end points of the oscilloscope trace are in the same horizontal level.

   (b) x feedthrough linearity compensation: Adjust R_{10} and/or R_{11} such that the U-shaped oscilloscope trace turns into a rotated sine-shaped trace.

   (c) Repeat step 3a until the rotated sine-shaped trace turns to a regular horizontal sine-shaped trace.

4. (a) Gain error: Set up the equipment as in Fig. 3.6, Sec. 3.2, to crossplot gain errors. Connect x to the ± 10 V, 5 Hz sine-wave generator and y to the $+10$ V dc reference voltage. Adjust R_{17} such that the two end points of the oscilloscope trace are in the same horizontal level.

   (b) Gain linearity compensation: Adjust R_8 and/or R_9 such that the U-shaped oscilloscope trace turns into a rotated sine-shaped trace.

   (c) Repeat step 4a until the rotated sine-shaped trace turns to a regular horizontal sine-shaped trace.

5. Repeat steps 1 through 4 as necessary.

6. Total error measurement: The equipment setup is similar to that in step 4. Interchange the x and y connections so that four error curves are crossplotted with the following four combinations of inputs, respectively:

   (a) $x = 10 \sin 2\pi 5t$, $y = +10$ V dc

   (b) $x = 10 \sin 2\pi 5t$, $y = -10$ V dc

   (c) $x = +10$ V dc, $y = 10 \sin 2\pi 5t$

   (d) $x = -10$ V dc, $y = 10 \sin 2\pi 5t$

   None of the four error curves can exceed the specified maximum total error limit.

3.4 Converting Any One-Quadrant Multiplier to Four-Quadrant Operation

Any one-quadrant multiplier can be converted to operate in all four quadrants by using three external operational amplifiers, as shown in Fig. 3.21. For the v_x input summing amplifier, we have

$$x = +10 - v_x$$

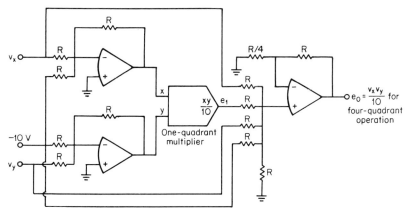

Fig. 3.21 Converting any one-quadrant multiplier to four-quadrant operation.

Similarly, v_y and y are given by

$$y = +10 - v_y$$

The output of the one-quadrant multiplier becomes

$$e_1 = \frac{xy}{10}$$

$$= 10 - v_x - v_y + \frac{v_x v_y}{10}$$

The output amplifier combines e_1, v_x, v_y, and -10 V, and its output voltage e_o is

$$e_o = e_1 + v_x + v_y - 10$$

$$= \frac{v_x v_y}{10}$$

v_x, v_y, and e_o can be any voltage between -10 and $+10$ V, while x, y, and e_1 are limited to the first quadrant between 0 and $+10$ V only. In the conversion of most one-quadrant multipliers, the three external amplifiers can be saved by making use of the input and output amplifiers internal to the one-quadrant multiplier. An illustration example can be found in Sec. 3.3.6 under log-antilog multiplier.

3.5 Analog Dividers

An analog divider provides an output voltage E_o as a function of two input voltages N and D such that

$$E_o = K \frac{N}{D}$$

108 Function Circuits

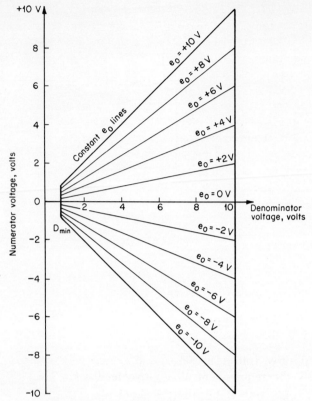

Fig. 3.22 Operating region for specified accuracy of analog dividers.

where K is a scaling constant. With standard ±15 V power supplies, K is often chosen as 10 in order to allow 10 V swing at the output and the inputs. Hence the transfer function is normally

$$E_o = 10 \frac{N}{D}$$

The denominator voltage D is always unipolar. Since D must not be zero or E_o will become infinity, manufacturers always specify a minimum denominator voltage D_{min} below which the divider is not guaranteed to operate within specifications. A high-performance divider operates satisfactorily at D_{min} as low as 10 mV, while a general-purpose low-cost divider may have unacceptable errors at D_{min} of 1 V. Because 10 V is the maximum rated output, the absolute value of numerator voltage must be less than that of denominator voltage. There are two main kinds of dividers, one-quadrant and two-quadrant. The numerator of a two-quadrant divider will accept bipolar voltages. The region of operation is within the trapezoid defined by the four heavy lines as shown in Fig. 3.22. For one-quadrant dividers, the numerator can only accept single-polarity voltages, and the operating region is half of the trapezoid, separated by the constant $e_o = 0$ V line. Some dividers are designed to accept only negative denominator voltages. Their operating region will be the mirror image from the vertical axis in Fig. 3.22.

3.5.1 Multiplier-inverted divider Two-quadrant analog division can be done by connecting any general-purpose four-quadrant multiplier in the feedback loop of an operational amplifier. Figure 3.23 illustrates the circuit connections of such an analog divider. Assuming the multiplier and the operational amplifier are ideal, the following two equations result:

$$v_o = \frac{xe_o}{10}$$

and

$$\frac{v_o}{R} = -\frac{z}{R}$$

Eliminate v_o from the above two equations; then

$$e_o = -10\frac{z}{x}$$

Note that x must be a positive voltage because z and e_o must be of opposite polarity to satisfy the requirement of negative feedback. The divider error depends on the accuracy of the multiplier and the offset voltage of the operational amplifier. If the multiplier has a multiplication error ϵ, v_o becomes

$$v_o = \frac{xe_o}{10} + \epsilon \tag{3-43}$$

Assuming that the input offset voltage of the operational amplifier is Z_o, then

$$\frac{v_o - Z_o}{R} = \frac{Z_o - z}{R} \tag{3-44}$$

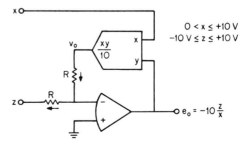

Fig. 3.23 A two-quadrant multiplier-inverted analog divider.

Eliminate v_o from Eqs. (3-43) and (3-44) and simplify:

$$e_o = -10\frac{z}{x} - \frac{10\epsilon}{x} + \frac{20Z_o}{x} \qquad (3\text{-}45)$$

The first term in Eq. (3-45) represents an ideal divider, while the second and third terms are error terms. Hence,

$$\text{Divider error} = -10\frac{\epsilon}{x} + \frac{20Z_o}{x}$$

Obviously, the divider error becomes excessively large for small values of x. A 10:1 denominator range is usually the practical limit. Special care should be taken not to use those multipliers, such as the quarter-square, whose errors increase with decreasing input signals.

Most modular multipliers in the market, and some integrated-circuit types, contain an output operational amplifier and thus can save the external amplifier and operate as a divider simply by variation of external pin connections.

3.5.2 Log-antilog divider Two-quadrant log-antilog dividers have recently been developed to provide excellent accuracy over a 60 dB dynamic range of denominators. Total error of 0.1 percent of full scale is achievable for denominators from 10 V down to 100 mV. The theory of operation is the same as that for the log-antilog multiplier discussed in Sec. 3.3.6. Let us first start with one-quadrant operation only. Rewrite Eq. (3-42), the transfer function for the log-antilog circuit in Fig. 3.19:

$$e_o = \frac{xy}{z}$$

There are two different ways to build a divider with this basic circuit. One way is to connect a +10 V dc reference voltage to z; the resulting multiplier, $e_o = xy/10$, is then inverted to a divider as discussed in the last section. This divider will have unacceptable errors at small denominator voltages comparable to those in Eq. (3-45). The other alternative is to set the +10 V dc reference voltage at x and use z as the denominator input and y as the numerator input. The transfer function, $e_o = 10y/z$, is that of a divider. Dividers thus connected are called log-antilog dividers and provide the best accuracy currently available.

To convert the one-quadrant divider to operate in two quadrants, level shifting is required. The complete circuit diagram is given in Fig. 3.24, where $-10 \text{ V} \leq y \leq +10 \text{ V}$, $0 < z \leq +10 \text{ V}$, and $e_o = 10y/x$. R_2, R_4, R_{14}, R_{15}, and R_{16} are used for level-shifting purposes. Instead of providing an expensive +10 V dc reference voltage, a 6.2 V temperature-compensated zener diode is called out. C_1/R_6, C_2/R_7, and C_3/R_{19} are phase compensation

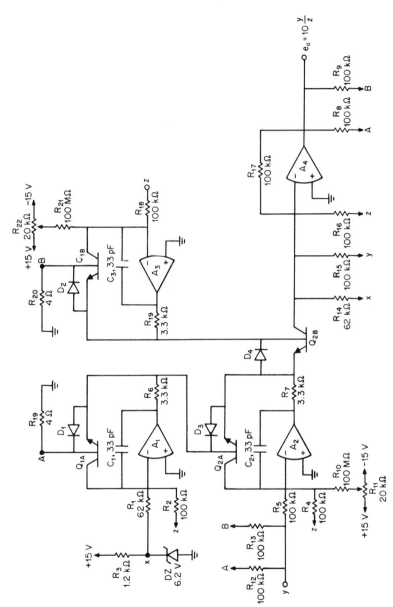

Fig. 3.24 A precision two-quadrant log-antilog divider.

111

networks to stabilize the three amplifiers, respectively. Q_{1A}/Q_{1B} and Q_{2A}/Q_{2B} must be matched pairs, and all four transistors must be at the same temperature. Diodes D_1, D_2, D_3, and D_4 are needed to protect the emitter-base junctions of the four respective logging transistors in case of accidental reverse bias. R_8/R_9, R_{12}/R_{13}, R_{19}, and R_{20} are provided for gain linearity and y input linearity compensations to achieve optimum accuracy. The input bias current and offset voltage of operational amplifiers A_2 and A_3 are extremely critical and sensitive at low denominator voltages; hence R_{10}/R_{11} and R_{21}/R_{22} are provided to null them.

Adjustment procedures:

1. Output offset: Ground y and x (short out zener diode DZ), set $z = +10$ V dc, and adjust R_4 and/or R_{18} for $e_o = 0$.
2. x input offset: Set $z = +10$ V dc and ground y; adjust R_1 and/or R_{14} such that $e_o = 0$.
3. (a) y input offset or y feedthrough: Set up the equipment as in Fig. 3.8a, Sec. 3.2, to crossplot the feedthrough on an oscilloscope. Ground x (short out DZ), and connect the $+10$ V dc reference to z and the ± 10 V, 5 Hz sine-wave generator to y and to the horizontal input of the oscilloscope. Adjust R_5 and/or R_{15} such that the two end points of the oscilloscope trace are in the same horizontal level.
   (b) y feedthrough linearity compensation: Adjust R_{12} and/or R_{13} such that the U-shaped oscilloscope trace turns into a rotated sine-shaped trace.
   (c) Repeat step 3a until the rotated sine-shaped trace turns to a regular horizontal sine-shaped trace.
4. (a) Gain error: Set up the equipment as in Fig. 3.6, Sec. 3.2, to crossplot gain errors. Connect z to the $+10$ V dc reference and y to the ± 10 V, 5 Hz sine-wave generator. Adjust R_{17} such that the two end points of the oscilloscope trace are in the same horizontal level.
   (b) Gain linearity compensation: Adjust R_8 and/or R_9 such that the U-shaped oscilloscope trace turns into a rotated sine-shaped trace.
   (c) Repeat step 4a until the rotated sine-shaped trace turns to a regular horizontal sine-shaped trace.
5. Repeat steps 1 through 4 as necessary.
6. Amplifier A_2 bias current and offset voltage adjustments: Set $z = +100$ mV dc or the expected minimum denominator voltage required in a specific application, and set $y = -z$. Note that the absolute values of y and z must be closely matched, preferably to within 0.05 percent. Adjust R_{11} such that $e_o = -10.000$ V $\pm$ 5 mV dc.
7. Amplifier A_3 bias current and offset voltage adjustments: Set $z = y = +100$ mV dc or the expected minimum denominator voltage. Adjust R_{22} such that $e_o = +10.000$ V $\pm$ 5 mV dc.

8. Repeat steps 6 and 7 as necessary.
9. Repeat steps 1 through 8 as necessary.

3.5.3 Offsetting a one-quadrant divider to accept bipolar numerator voltages

There are a few commercial one-quadrant dividers available that will accept only single-polarity voltages for both the denominator and numerator.

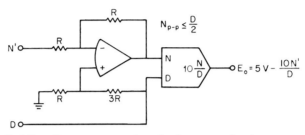

Fig. 3.25 Offsetting a one-quadrant divider to accept bipolar numerator voltages.

Even though the divider may be designed for primarily dc signals, it usually has sufficient bandwidth to be used as a dc-controlled variable-gain ac amplifier, i.e., the denominator will be a unipolar dc voltage, while the numerator will be a bipolar ac voltage. Significantly, using the divider as a variable-gain device produces gains greater than unity, which is not possible with general-purpose multipliers. The circuit shown in Fig. 3.25 uses one additional amplifier with the one-quadrant divider and produces analog division for ac numerators. The peak-to-peak voltage at N must be less than or equal to $D/2$. The maximum voltage at E_o will be 10 V p-p with an average value of 5 V dc. The output of the divider may be ac-coupled if desired, or the output may be offset back to zero by an additional output amplifier. Note that the output of the divider will be inverted with respect to the N' input.

3.6 Multifunction Converters

The multifunction converter provides a low-cost solution to many analog conversion requirements. More than just another multiplier/divider, it is capable of squaring, square rooting, squaring of ratios, and raising ratios to arbitrary powers with a very high degree of accuracy. The general transfer function is

$$e_o = v_y \left(\frac{v_z}{v_x}\right)^m$$

v_x, v_y, and v_z refer to input voltages. The exponent m is determined by the selection of two external resistors; 0.2 percent accuracy is achievable for m ranging from 0.2 to 5. With the addition of a few passive and active components, functions such as true root-mean-square computation; sine, cosine, and arctangent function generation; and vector sums can easily be implemented. A more detailed discussion of the applications of the multifunction converter is presented in Chapter 5. Because of its versatility and relatively simple circuitry, more and more applications are expected to be discovered in the near future.

3.6.1 Theory of operation The multifunction converter employs log-antilog techniques. The block diagram is given in Fig. 3.26. Basically it consists of four operational amplifiers, four logging transistors, and four resistors. Rewrite Eq. (3-6), the logarithmic equation for bipolar transistors:

$$v_{BE} = \frac{kT}{q} \ln \frac{i_c}{I_S} \qquad (3\text{-}46)$$

Apply Eq. (3-46) to logging transistors Q_z and Q_x:

$$v_B - v_1 = \frac{kT}{q} \ln \frac{i_z}{I_{Sz}} \qquad (3\text{-}47)$$

and

$$0 - v_1 = \frac{kT}{q} \ln \frac{i_x}{I_{Sx}} \qquad (3\text{-}48)$$

I_{Sz} and I_{Sx} are the emitter saturation currents for transistors Q_z and Q_x, respectively. Subtract Eq. (3-48) from (3-47):

$$v_B = \frac{kT}{q} \ln \frac{i_z I_{Sx}}{i_x I_{Sz}}$$

If transistors Q_z and Q_x are on the same monolithic chip and their characteristics are closely matched, it can be assumed that $I_{Sx} = I_{Sz}$. Then

$$v_B = \frac{kT}{q} \ln \frac{i_z}{i_x} \qquad (3\text{-}49)$$

Similarly, it can be derived for logging transistors Q_y and Q_o that

$$v_C = \frac{kT}{q} \ln \frac{i_o}{i_y} \qquad (3\text{-}50)$$

Consider the first case, where v_A, v_B, and v_C are shorted together to make $m = 1$. Since $v_B = v_C$, equate Eqs. (3-49) and (3-50):

$$\frac{i_z}{i_x} = \frac{i_o}{i_y} \qquad (3\text{-}51)$$

Multipliers, Dividers, and Multifunction Converters 115

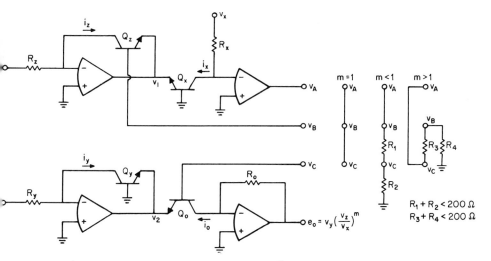

3.26 Block diagram of the multifunction converter, $e_o = v_y \left(\dfrac{v_z}{v_x} \right)^m$

Substitute $i_z = v_z/R_z$, $i_x = v_x/R_x$, $i_y = v_y/R_y$, and $i_o = e_o/R_o$ into Eq. (3-51) and simplify.

$$e_o = \left(v_y \dfrac{v_z}{v_x} \right) \left(\dfrac{R_x R_o}{R_y R_z} \right)$$

If we make $R_x = R_y = R_z = R_o$, then

$$e_o = v_y \dfrac{v_z}{v_x} \qquad (3\text{-}52)$$

For the second case, where $m < 1$, short out v_A to v_B and use two resistors R_1 and R_2 to drop the voltage at v_B down to v_C as shown in Fig. 3.26. Then

$$v_C = m v_B \quad \text{where } m = \dfrac{R_2}{R_1 + R_2} \qquad (3\text{-}53)$$

Substitute Eq. (3-53) into (3-50), then equate the result with (3-49) and simplify.

$$i_o = i_y \left(\dfrac{i_z}{i_x} \right)^m$$

Converting i_o, i_y, i_z, and i_x to e_o, v_y, v_z, and v_x, respectively, we have

$$e_o = v_y \left(\dfrac{v_z}{v_x} \right)^m \dfrac{R_o}{R_y} \left(\dfrac{R_x}{R_z} \right)^m$$

116 Function Circuits

Again making $R_o = R_y = R_x = R_z$, this simplifies to

$$e_o = v_y \left(\frac{v_z}{v_x}\right)^m \tag{3-54}$$

The third case, where $m > 1$, can be derived similarly by shorting v_A to v_C and dropping the voltage at v_C down to v_B by two resistors, R_3 and R_4, as shown in Fig. 3.26. Then

$$e_o = v_y \left(\frac{v_z}{v_x}\right)^m \quad \text{where } m = \frac{R_3 + R_4}{R_4} \tag{3-55}$$

3.6.2 Practical multifunction converter circuit To design a practical multifunction converter circuit, the block diagram in Fig. 3.26 needs to be frequency-compensated and reverse-bias-protected as shown in Fig. 3.27. RC circuits of 3.3 kΩ and 22 pF are used to stabilize amplifiers A_z and A_y. A fairly large capacitor, 0.0047 μF, is needed to prevent amplifier A_x from oscillating. Diodes are used to protect the base-emitter junctions of the logging transistors in case they are accidentally reverse-biased. The input resistor at v_y is intentionally designed to be 10 percent below the other three 100 kΩ resistors so that a variable resistor of 25 kΩ can be connected in series with the 90 kΩ resistor to null out the tolerances in the three 100 kΩ resistors and the errors in the logging transistors. The external resistors used to set the exponent m should be kept below 200 Ω in order to prevent errors due to base current.

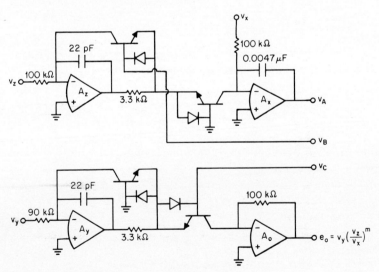

Fig. 3.27 A practical multifunction converter circuit.

3.6.3 Multiplication and division

The multifunction converter may be used as a precision one-quadrant multiplier or divider by shorting out v_A, v_B, and $v_C (m = 1)$. To use it as a multiplier, apply a $+10$ V dc reference voltage at v_x, and to use it as a divider, apply a $+10$ V dc reference at v_y, as in Fig. 3.28. In most cases, the $+15$ V dc power-supply voltage is stable

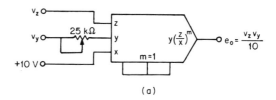

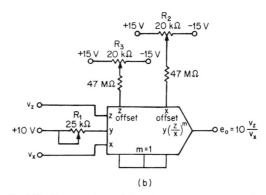

Fig. 3.28 Connect the multifunction converter as a precision (a) one-quadrant multiplier or (b) one-quadrant divider.

enough to serve as the reference, and to save the separate $+10$ V reference, simply connect a 50 kΩ resistor between the supply and the respective input terminal.

In multiplier applications, only one adjustment is needed to achieve less than 0.25 percent of full-scale error for any combination of v_z and v_y from $+10$ mV to $+10$ V. Simply set $v_z = v_y = +10.000$ V $\pm$ 1 mV dc and adjust the 25 kΩ potentiometer such that $e_o = +10.000$ V $\pm$ 1 mV dc. From -25 to $+85°$C, the multiplier drifts about 1 mV/°C.

As a divider, the multifunction converter is far superior to the best available multiplier-inverted dividers. For a denominator range of 100:1, total error is typically less than 0.25 percent. At low denominator voltages the divider output error becomes extremely sensitive to the input bias currents and offset voltages of the v_z and v_x input amplifiers. Unless high-performance operational amplifiers are used for A_z and A_x in Fig. 3.27, offset adjustments are normally required. In Fig. 3.28b, the z offset and x

118 Function Circuits

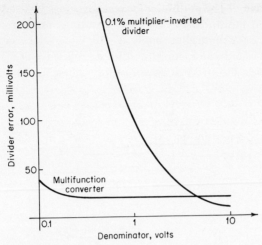

Fig. 3.29 Comparison of divider error curves for the multifunction converter and a 0.1 percent multiplier-inverted divider.

offset terminals of the multifunction converter correspond to the summing junctions of A_z and A_x in Fig. 3.27, respectively. Perform the following adjustment procedures:

1. Set R_1 such that with $v_z = v_x = +10.000 \text{ V} \pm 1$ mV dc, $e_o = +10.000$ V ± 1 mV dc.
2. Set R_2 such that with $v_z = v_x = +100$ mV ± 0.1 mV dc, $e_o = +10.000$ V ± 1 mV dc.

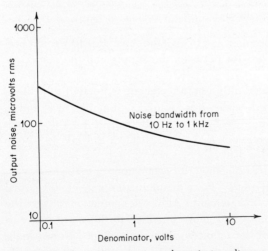

Fig. 3.30 Divider output noise versus denominator voltage.

3. Set R_3 such that with $v_z = +10$ mV $\pm$ 0.1 mV dc and with $v_x = +100$ mV $\pm$ 0.1 mV dc, $e_o = +1.000$ V $\pm$ 0.2 mV dc.
4. Repeat steps 1 through 3 if necessary.

Most commercially available multifunction converters are specified at a maximum error of 0.5 percent of full scale for v_z in the range between 0.01 and 10 V and v_x between 0.1 and 10 V. As compared to a conventional 0.1 percent multiplier-inverted divider whose error is inversely proportional to its denominator voltage, the multifunction divider is capable of holding

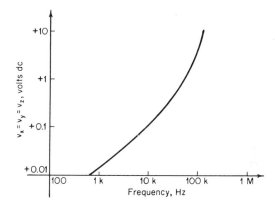

Fig. 3.31 Small-signal (10 percent of dc level)–3 dB bandwidth.

its error curve fairly constant down to 0.1 V, as illustrated in Fig. 3.29. A plot of output noise versus denominator voltage is given in Fig. 3.30. Note that the noise does not go up by a factor of 100 as in multiplier-inverted dividers, as the denominator voltage goes down by 100:1.

Full-power output frequency bandwidth with a ± 4 V ac signal superimposed on the +6 V dc level is typically 60 kHz in the multiply mode with 741-type amplifiers. In the divide mode, the bandwidth increases with increased denominator voltages v_x. A close approximation is 6 kHz times v_x, e.g., with v_x at +5 V dc, the bandwidth is 30 kHz. As illustrated in Fig. 3.31, the small-signal -3 dB bandwidth for all three inputs v_x, v_y, and v_z depends on the dc level on which the 10 percent ac signal is superimposed. To achieve higher frequency response, it should be biased at the maximum dc level allowed by the specific applications.

3.6.4 Exponential functions The multifunction converter may be used as a precision exponentiator over a range of exponents from 0.2 to 5. The curves given in Fig. 3.32 illustrate the effect of varying the exponent m in

$$e_o = 10\left(\frac{v_z}{10}\right)^m$$

Figure 3.33a and depicts the interconnections for $m < 1$ and $m > 1$, respectively. The 25 kΩ potentiometer should be adjusted such that with $v_z = +10.000$ V $\pm$ 1 mV dc, $e_o = +10.000$ V $\pm$ 1 mV dc. If it is required to adjust m continuously passing through unity, a potentiometer R_A together with two fixed resistors R_B connected as in Fig. 3.33c may be used. If the resistance value on one side of the potentiometer wiper is KR_A, then the value on the other side will be $(1 - K)R_A$, where $0 \leq K \leq 1$. From Fig. 3.33c,

$$v_B = \frac{R_B}{R_B + KR_A} v_A$$

and

$$v_C = \frac{R_B}{R_B + (1 - K)R_A} v_A$$

Following the derivation procedures given in Sec. 3.6.1 and starting with Eqs. (3–49) and (3–50), it can be proved that

$$m = \frac{R_B + KR_A}{R_B + (1 - K)R_A} \tag{3–56}$$

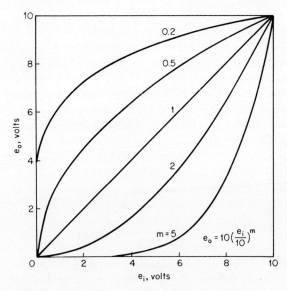

Fig. 3.32 Exponentiator transfer characteristics.

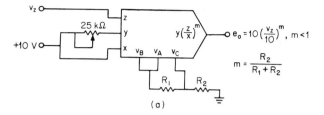

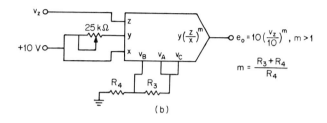

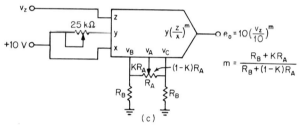

Fig. 3.33 Exponentiators, $e_o = 10 \left(\dfrac{v_z}{10}\right)^m$, (a) $m < 1$, (b) $m > 1$, (c) $\dfrac{R_B}{R_A + R_B} < m < \dfrac{R_A + R_B}{R_B}$.

Considering the extreme case when the wiper of the potentiometer is on the left end, where $K = 0$, Eq. (3-56) becomes $m = R_B/(R_A + R_B)$. When the wiper is at the middle, where $K = 0.5$, $m = 1$. If the wiper is moved to the right extreme such that $K = 1$, then $m = (R_A + R_B)/R_B$. For instance, by making $R_A = 500\ \Omega$ and $R_B = 125\ \Omega$, the exponent m can be swept from 0.2, through unity, up to 5.

3.7 Square Rooters

Square rooters are commonly used blocks that extract the square root of an input voltage. General-purpose analog multipliers, precision dividers, and multifunction converters can all be used for square-rooting, but the accuracies differ widely among the three approaches. The frequency response of a square rooter is difficult to describe because the square-rooting function is very nonlinear. The gain varies significantly with the input level. For an input voltage z, the output e_o of a square rooter is given by

$$e_o = \sqrt{10z}$$

The small-signal gain is

$$\frac{de_o}{dz} = \frac{5}{\sqrt{10z}}$$

So the small-signal gain varies from 50 to 0.5 as the input varies from 1 mV to +10 V. The small-signal frequency response depends on the gain — the response is much slower for low-level inputs than for high-level inputs. The phase shift is much greater for low-level inputs. The step response also depends upon the output level. Since the gain is low, and the response faster, at the high output levels, the response to a positive-going step is very rapid. But a step from a high output level to a low output level means that the unit is responding with high gain, and therefore slow response.

37.1 Multiplier-inverted square rooter Any general-purpose multiplier can be connected in the feedback loop of an operational amplifier to perform square root functions. The circuit diagram is shown in Fig. 3.34a.

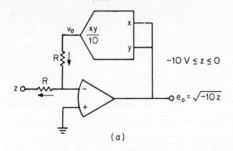

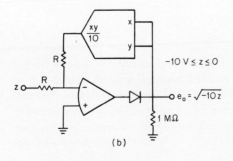

Fig. 3.34 (a) Multiplier-inverted square-rooter. (b) Protection network included to stop circuit latchup.

Assuming an ideal multiplier and an ideal operational amplifier, the following two equations result:

$$v_o = \frac{e_o^2}{10} \qquad (3\text{-}57)$$

and

$$\frac{v_o}{R} = -\frac{z}{R} \qquad (3\text{-}58)$$

Eliminating v_o, we have

$$e_o = \sqrt{-10z} \qquad (3\text{-}59)$$

Obviously z must be a negative voltage, and e_o is always positive. The circuit will sometimes latch up because of instantaneous positive feedback caused by noise voltages. Hence a protection diode is usually required, as illustrated in Fig. 3.34b. The 1 MΩ resistor is needed to turn on the diode if the x and y input impedances of the multiplier are extremely high. Most modular multipliers, and some IC types, contain an output operational amplifier and thus can be operated as square rooters simply by varying external pin connections.

If the multiplier has a multiplication error ϵ and the operational amplifier has an input offset Z_o, then Eqs. (3-57) and (3-58) become

$$v_o = \frac{e_o^2}{10} + \epsilon \qquad (3\text{-}60)$$

and

$$\frac{v_o - Z_o}{R} = \frac{Z_o - z}{R} \qquad (3\text{-}61)$$

From Eqs. (3-60) and (3-61),

$$e_o = \sqrt{-10z + 10(2Z_o - \epsilon)} \qquad (3\text{-}62)$$

It can be determined from Eq. (3-62) that errors become troublesome for small values of z. It is only practical to operate over input voltages of 10 V to 100 mV.

3.7.2 Divider-converted square rooter As we have discussed in the preceding sections, any general-purpose multiplier can be connected to perform square-root functions, but the accuracy is reasonable only for a very limited dynamic range. To achieve high accuracy for small values of input voltages, a precision divider, such as the log-antilog, should be converted to a square rooter. Figure 3.35 illustrates the circuit connections. The diode is used to prevent circuit latch-up. A resistor should be connected

124 Function Circuits

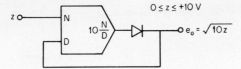

Fig. 3.35 Divider-converted square rooter.

between e_o and ground to turn on the diode if the D input has extremely high input impedance. Depending on the accuracy of the divider, a square-root accuracy of less than 20 mV is readily achievable over input voltages of 10 V to 1 mV.

3.7.3 Use the multifunction converter as a precision square rooter The multifunction converter can be used, through proper external connections, as a precision square rooter. The square rooter is a special case of the exponentiator where $m = 0.5$. It can be implemented either by using two

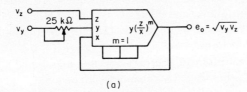

(a)

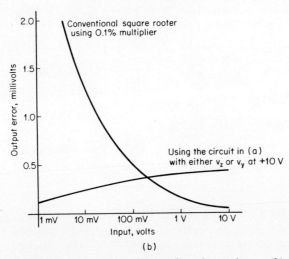

(b)

Fig. 3.36 (a) Square rooter of the product of two voltages. (b) Comparison of error curves of a conventional square rooter, $e_o = \sqrt{10e_i}$, using a 0.1 percent multiplier and the square rooter in (a) with a +10 V reference at either v_y or v_z.

precision resistors of equal value, as in Fig. 3.33a, or, preferably, by feeding back the output to v_x. Making $m = 1$,

$$e_o = v_y \frac{v_z}{v_x}$$

If e_o and v_x are connected, the circuit in Fig. 3.36a takes the square root of the product of two input voltages, i.e.,

$$e_o^2 = v_y v_z \quad \text{or} \quad e_o = \sqrt{v_y v_z}$$

Putting a +10 V reference voltage at either v_y or v_z makes this a conventional square rooter, $e_o = \sqrt{10 e_i}$. This approach provides excellent accuracy over a 10 V to 1 mV input range. Below 1 mV, noise and drift become excessive. Square rooters using conventional precision multipliers have error curves going up exponentially with low input voltages, while the error of this square rooter actually goes down slightly. Comparison curves are given in Fig. 3.36b.

REFERENCES

1. G. E. Tobey, Analog Modules Multiply User's Options, *Electron. Products*, Feb. 19, 1973.
2. H. Schmid, Build a Precision Pulse-Time Multiplier, *Electron. Des.*, April 27, 1972.
3. B. Gilbert, A High Performance Monolithic Multiplier Using Active Feedback, *IEEE J. Solid-State Circ.*, vol. SC-9, no. 6, December 1974.
4. B. Gilbert, A High Accuracy Analog Multiplier, *ISSCC*, Feb. 14, 1974.
5. L. Counts, Reduce Multiplier Errors by Up to an Order of Magnitude, *EDN*, March 20, 1974.
6. H. K. Henson, Electronic Analog Multiplier, *United States Patent* 3,805,092, April 16, 1974.

4
RMS-TO-DC CONVERSION

Precise measurement of dc signals is not trivial; however, only two quantities will fully describe the signal: polarity and magnitude. A complete description of an ac signal is much more complex and requires knowing whether the signal is periodic, and if so, what the period or repetition rate is; what the signal's shape is; what its peak-to-peak value is; and finally its magnitude or rms value. All this information can be contained in an infinite series such as a Fourier series, and a table of the polarity, magnitude, and frequency of the most significant terms of the signal's Fourier series could be used as the measure of an ac signal. Such a complete description of the signal is not required in most engineering applications, however, and a suitable description can be obtained with an oscilloscope and an ac voltmeter or ammeter. In many cases, sufficient information about the signal's shape and frequency is already known, and the signal's magnitude is all that is required. This chapter deals with the circuits used to develop a dc voltage which is proportional to the rms value of an ac signal.

4.1 General AC-to-DC Conversion

There are three amplitude values of an ac signal which are commonly converted to a dc voltage:

1. Peak (or peak-to-peak) amplitude E_p.
2. Average value, which is actually the average absolute value, since the average value of symmetrical waveforms is zero. The average value of an ac waveform is equal to the dc level which will transfer the same charge per unit time, and mathematically the average absolute value is

$$E_{\text{ave}} = \frac{1}{t_o} \int_0^{t_o} |e(t)| \, dt$$

where t_o is the interval of time over which the waveform's value is of interest.
3. Root-mean-square or rms value, which is a measure of the signal's energy content. The mathematical expression for the rms value over the interval of time t_o is

$$E_{\text{rms}} = \sqrt{\frac{1}{t_o} \int_0^{t_o} e^2(t) | dt}$$

In most practical applications a continuous measure of the rms value is desired rather than the rms value over a specific interval of time. Therefore, an important but subtle approximation is that the expression under the radical is the average value of $e^2(t)$.

The mathematical expression is also shortened by this approximation:

$$E_{\text{rms}} = \sqrt{\overline{e^2(t)}}$$

where the bar (———) denotes average value.

For a given waveform the three dc values, E_p, E_{ave}, and E_{rms}, can be considered equivalent measures of magnitude, since they have fixed ratios for a fixed waveform. The ratio of the peak value to the rms value is commonly called the *crest factor*, and the ratio of the rms value to the average value is commonly called the *form factor*.

$$\text{Crest factor} = K_C = \frac{E_p}{E_{\text{rms}}}$$

$$\text{Form factor} = K_F = \frac{E_{\text{rms}}}{E_{\text{ave}}}$$

4.1.1 Peak-responding ac-to-dc converters

A precision peak-responding ac-to-dc converter is shown in Fig. 4.1. Two operational amplifiers are used; one provides very high input impedance, and the other allows output loading without degradation of the dc output voltage. The holding capacitor is shunted by a resistor to provide a running peak measurement, which is most useful with periodic signals. The RC time constant should be long compared with the period of the input signal to provide a steady reading of

128 Function Circuits

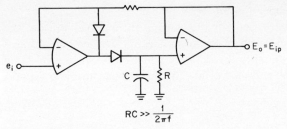

$$RC \gg \frac{1}{2\pi f}$$

Fig. 4.1 Precision peak-responding ac-to-dc converter.

the input signal's peak value. Peak-responding ac measurements are most common with pulse train types of signals, where the peak value may be of most interest. However, the output of a peak-responding converter may be scaled to provide a dc output proportional to the rms value of the input, assuming a sinusoidal waveform. In which case, the crest factor is 1.414, and the output is then attenuated by the factor 0.707 to provide the equivalent rms value. Low-cost and particularly wide bandwidth ac-to-dc converters often measure the peak value and then convert to rms value, since a very simple but very wideband peak detector can be built, as in Fig. 4.2, that will provide satisfactory accuracy for many wide-bandwidth measurements.

$$E_o = \frac{E_{ip}}{K_C} = E_{irms}$$
$$(E_{ip} \gg V_D)$$

$$(R_1 + R_2)C \gg \frac{1}{2\pi f}, \quad \frac{R_1 + R_2}{R_2} = K_C = \text{Crest factor}$$

Fig. 4.2 Wide-bandwidth rms converter utilizing low-cost peak-responding circuitry.

4.1.2 Average-responding ac-to-dc converters A very common technique for generating a dc voltage proportional to the rms value of an ac signal is to use an average-responding ac-to-dc converter. A suitable precision ac-to-dc converter is shown in Fig. 4.3. The output of this circuit is a running average of the input, and this can be easily scaled to provide a dc output proportional to the rms value for a given waveform. With sinusoidal inputs the form factor is 1.11072, and if it is assumed that the input is purely sinusoidal, the output amplifier's gain should then be 1.11072 in order to provide an equivalent rms value.

While the circuits in both Figs. 4.1 and 4.3 provide a precise measure of their respective values, peak or average, the average-responding circuit in Fig. 4.3 usually provides the best accuracy when the rms value of the input is actually desired. The averaging ac-to-dc converter responds to the total signal, not to only one polarity of the signal's peak, and is therefore less distortion- and noise-sensitive than the peak-responding ac-to-dc converter. However, even the averaging ac-to-dc converter only provides

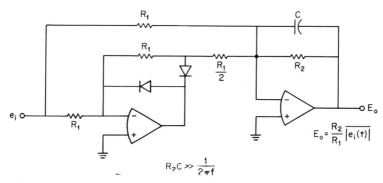

Fig. 4.3 Precision average-responding ac-to-dc converter.

good accuracy with relatively pure sine-wave inputs (if, of course the form factor for sine waves was used as the circuit's scale factor).

4.1.3 Why use the rms value? In all but a few cases it is the energy content of a signal that we are most interested in, not the signal's peak or average value, and the rms value is a measure of the signal's energy content. The rms value of ac waveforms is so widely accepted that, unless otherwise indicated, most engineers assume that the rms value is meant when ac voltages are specified. Rms values are particularly important in the case of noise measurements, since noise levels are referenced to power. The rms value is also important in statistical studies, since the standard deviation of a stationary random process with zero mean is the rms value of the process. Further, if uncorrelated or orthogonal quantities are summed, the rms value of the sum is equal to the square root of the sum of the squares of the individual rms values.

While the peak- and average-responding ac-to-dc converters discussed can be scaled to provide a dc output equal to the rms value of the input for a given input waveform by application of the appropriate crest factor and form factor, the full properties of the rms value will not be realized. Suppose an average-responding ac-to-dc converter is scaled to provide the rms value for pure sine waves and is then used to measure the output of a good sine-wave oscillator with a maximum distortion of 1 percent. Depending on the frequency and the phase of the distortion, up to 0.33 per-

TABLE 4.1 Factors for Converting Several Frequently Encountered Waveforms to True RMS, for Common Measuring Techniques.

Waveform	Average (rms calibrated)	True average	Peak (rms calibrated)	True peak
Dc level	$\frac{2\sqrt{2}}{\pi}$	1	$\sqrt{2}$	1
Sine	1	$\frac{\pi}{2\sqrt{2}}$	1	$\frac{1}{\sqrt{2}}$
Half-wave rectified sine	$\sqrt{2}$	$\frac{\pi}{2}$	$\frac{1}{\sqrt{2}}$	$\frac{1}{2}$
Triangle	$\frac{4}{\pi}\sqrt{\frac{2}{3}}$	$\frac{2}{\sqrt{3}}$	$\sqrt{\frac{2}{3}}$	$\frac{1}{\sqrt{3}}$
Square	$\frac{2\sqrt{2}}{\pi}$	1	$\sqrt{2}$	1
Pulse	$\frac{2\sqrt{2}}{\pi}\sqrt{\frac{P}{W}}$	$\sqrt{\frac{P}{W}}$	$\sqrt{\frac{2W}{P}}$	$\sqrt{\frac{W}{P}}$
White noise	$\frac{2}{\sqrt{\pi}}$	$\sqrt{\frac{\pi}{2}}$	—	—

cent error can result with an average-responding measurement; an rms-responding measurement would be accurate independent of the frequency or the phase of the distortion, since it measures the total energy content independent of the time relationships. A peak-responding ac-to-dc converter has an even greater potential error, up to 1 percent error for 1 percent distortion. Even relative ac measurements, such as an amplifier's gain, can be in error as a result of small distortions in the input signal, since the amplifier under test may introduce phase shift between the input fundamental and the distortion. In fact, 1 percent distortion in the input sine wave may produce as much as 0.33 percent error with an average-responding measurement, depending on the phase shift introduced by the amplifier being tested. Again, if an rms-responding measurement were made, no error would result from the distortion, since the rms measurement is of the total energy content, independent of phase relationships.

For nonsinusoidal waveforms the errors which result with peak- or average-responding ac-to-dc converters are even worse. For example, when measuring a square wave which is symmetrical about zero, an average-responding converter calibrated to equal the rms input with sine waves will have an output error of +11 percent, and a peak-responding converter calibrated to equal the rms input with sine waves will have an output error −30 percent. Table 4.1 shows the conversion factors between the output of the common ac-to-dc converters and the true rms value for several commonly encountered waveforms. It is apparent from the conversion factors required that only a true rms-responding measurement is capable of providing an accurate output for anything other than pure sine waves.

4.2 Computing RMS Converters[1,2]

By means of analog computation a signal can be electrically processed through the mathematical operations required to derive its rms value. Straightforward computation of a signal's rms value is shown in Fig. 4.4a. The low-pass filter provides a running average of the input squared, and the square root of its output provides a continuous conversion of the input signal to a dc level corresponding to the rms value of the input. The negative square root is used in Fig. 4.4a, since the square-root connection of most multiplier/dividers inherently provides an inversion (see Chapter 3). While the direct computation of the rms value in Fig. 4.4a is easy

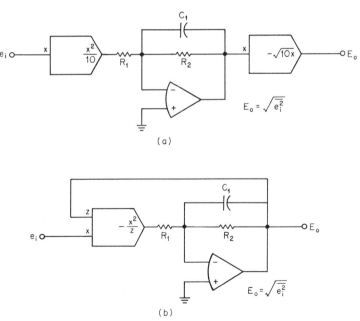

Fig. 4.4 Rms-to-dc conversion can be realized by (a) direct analog computation or (b) implicit computation.

to understand and can provide accuracies of the order of 0.5 percent, it is not the best configuration for a computing rms converter. An implicit method of computing a signal's rms value is shown in Fig. 4.4b; this is actually easier to implement if the log-antilog techniques discussed in Chapters 2 and 3 are used.

The differential equations for both the direct and the implicit rms converters can be easily derived by summing the currents at the inverting input of the operational amplifier used in the low-pass filter. The currents at the summing junction of the direct rms converter in Fig. 4.4a are

$$\frac{e_i^2}{10R_1} - \frac{e_o^2}{10R_2} - \frac{C_1}{10}\frac{de_o^2}{dt} = 0$$

and for the implicit rms converter in Fig. 4.4b they are

$$-\frac{e_i^2}{e_o R_1} + \frac{e_o}{R_2} + C_1 \frac{de_o}{dt} = 0$$

By using the identity

$$\frac{de_o^2}{dt} = 2e_o \frac{de_o}{dt}$$

and rearranging the terms, we obtain the two differential equations

$$e_o^2 + R_2 C_1 \frac{de_o^2}{dt} = \frac{R_2}{R_1} e_i^2 \qquad \text{direct rms conversion}$$

and

$$e_o^2 + \frac{R_2 C_1}{2} \frac{de_o^2}{dt} = \frac{R_2}{R_1} e_i^2 \qquad \text{implicit rms conversion}$$

Note that the only difference between the two differential equations is that the direct-converter time constant is $R_2 C_1$, while the time constant for the implicit converter is only $R_2 C_1/2$. The theoretical and practical considerations of an implicit-computing rms converter will be considered in the remainder of this section, but the results are general and can be applied to any rms converter if the appropriate time constant is used.

4.2.1 Implicit-computing rms converter

The implicit rms converter can be realized by using four logging transistors as in the log-antilog multiplier/divider discussed in Chapter 3. The base-emitter voltages of the logging transistors in the implicit rms converter in Fig. 4.5 are constrained to be

$$v_{BE1} + v_{BE3} = v_{BE2} + v_{BE4}$$

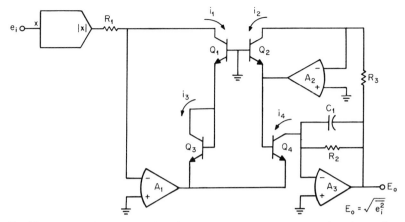

Fig. 4.5 Basic implementation of implicit computing rms converter utilizes log-antilog technique.

According to Eq. (2-1) in Chapter 2, the base-emitter voltages are logarithmically related to their collector current, resulting in the relationship

$$\frac{kT_1}{q}\ln\frac{i_1}{I_{S1}} + \frac{kT_3}{q}\ln\frac{i_3}{I_{S3}} = \frac{kT_2}{q}\ln\frac{i_2}{I_{S2}} + \frac{kT_4}{q}\ln\frac{i_4}{I_{S4}}$$

Assuming that all the transistors are at the same temperature, this expression can be simplified to

$$i_4 = \frac{I_{S2}I_{S4}}{I_{S1}I_{S3}}\frac{i_1 i_3}{i_2} \qquad (4\text{-}1)$$

Since the logging transistors are unipolar devices, the absolute-value circuit is necessary to provide a unipolar signal current. If amplifier A_1 is assumed to have negligible offset voltage and input bias current, the collector current of logging transistor Q_1 is

$$i_1 = \frac{|e_i|}{R_1}$$

and then the collector current of logging transistor Q_3 is

$$i_3 = \frac{\alpha_3}{\alpha_1}\frac{|e_i|}{R_1}$$

where α_1 and α_3 are the common-base current gains of Q_1 and Q_3, respectively. Assuming that operational amplifiers A_2 and A_3 also have negligible offset voltage and input bias current, the relationships for i_2 and i_4 can also be written

$$i_2 = \frac{e_o}{R_3}$$

and

$$i_4 = \frac{e_o}{R_2} + C_1 \frac{de_o}{dt}$$

Combining these expressions for the logging transistor collector currents with Eq. (4–1) yields the differential equation

$$\frac{I_{S2}I_{S4}}{I_{S1}I_{S3}} \frac{\alpha_3}{\alpha_1} \frac{|e_i|^2}{R_1^2} = \frac{e_o^2}{R_2 R_3} + \frac{e_o C_1}{R_3} \frac{de_o}{dt}$$

Combining this with the identities

$$|e_i|^2 = e_i^2$$

and

$$2e_o \frac{de_o}{dt} = \frac{de_o^2}{dt}$$

and rearranging, we obtain

$$e_o^2 + \frac{R_2 C_1}{2} \frac{de_o^2}{dt} = \frac{I_{S2}I_{S4}}{I_{S1}I_{S3}} \frac{\alpha_3}{\alpha_1} \frac{R_2 R_3}{R_1^2} e_i^2$$

In the complex frequency domain s the transfer function for this differential equation is

$$E_o^2(s) = \frac{I_{S2}I_{S4}}{I_{S1}I_{S3}} \frac{\alpha_3}{\alpha_1} \frac{R_2 R_3}{R_1^2} \frac{E_i^2(s)}{1 + sR_2 C_1/2} \tag{4-2}$$

For sinusoidal inputs

$$e_i(t) = \sqrt{2} E_{irms} \cos \omega t$$

and the squared input is

$$e_i^2(t) = E_{irms}^2 (1 + \cos 2\omega t) \tag{4-3}$$

If the frequency of the input ω is much greater than the inverse of the transfer-function time constant $R_2 C_1/2$, or if the input is a dc level, the steady-state transfer function simplifies to

$$E_o = \sqrt{\frac{I_{S2}I_{S4}}{I_{S1}I_{S3}} \frac{\alpha_3}{\alpha_1} \frac{R_2 R_3}{R_1^2}} E_{irms} \tag{4-4}$$

From this expression it can be seen that with an ideal absolute-value circuit and ideal operational amplifiers, only a gain error exists in the computing converter's midband rms-to-dc conversion. This gain error can easily be compensated by adjusting one of the resistors to force

$$\frac{I_{S2}I_{S4}}{I_{S1}I_{S3}} \frac{\alpha_3}{\alpha_1} \frac{R_2 R_3}{R_1^2} = 1$$

While any of the resistors could be adjusted, adjusting R_3 is recommended, since R_1 potentially has high-frequency currents in it and its frequency response could be impaired by potentiometers; and, possibly more importantly, in a practical absolute-value circuit such as the one used in Fig. 4.6, R_1 may be a combination of resistors. R_2 should not be adjusted either, since its value in conjunction with the capacitor determines the low-frequency cutoff and the step response.

The gain error will drift with temperature and time because of resistor drift and the temperature dependence of the current gains and the saturation currents. The overall drift at the output is approximately

$$\Delta E_o = \frac{E_{irms}}{2} \Delta \left(\frac{I_{S2}I_{S4}}{I_{S1}I_{S3}} \frac{\alpha_3}{\alpha_1} \frac{R_2 R_3}{R_1^2} \right)$$

By using precision resistors and monolithic pairs of transistors, grouped Q_1-Q_2 and Q_3-Q_4, the drift with temperature can be held to about 70 ppm/°C, or in terms of output voltage drift, $70 E_{irms}$ μV/°C. Input offset voltage and bias current in the operational amplifiers will produce additional errors, an output offset voltage and a dc reversal error. The dc reversal error is defined as the change in the output due to reversing the polarity of a dc input level. Even without adjustment of their offset voltages, moderately high performance operational amplifiers will provide an output offset of only about 5 mV, which is only 0.05 percent of full-scale error with a typical 10 V full scale. The dc reversal error in the computing rms converter is due to offsets at the input of A_1 and A_4, and also due to mismatch in the absolute-value circuit resistors. To obtain the ultimate performance, the offset voltage of both A_1 and A_4 should be nulled. Also, the symmetry of the absolute-value converter should be adjusted; this is easily done by putting a 500 Ω potentiometer in series with R_4 and R_5, with its slider connected to the summing junction of A_4. If only ac-coupled signals are to be measured, the dc reversal error will probably be insignificant, since it introduces an error in a square root of the sum of the squares fashion:

$$I_{1rms} = \frac{1}{R_1} \sqrt{E_{irms}^2 + \epsilon_{dcrev}^2} \qquad \text{ac-coupled input}$$

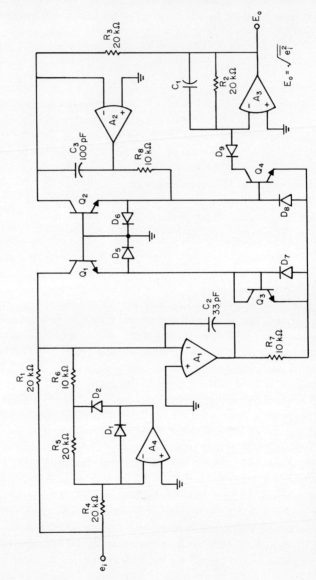

Fig. 4.6 Complete computing rms-to-dc converter utilizes log-antilog technique.

136

where ϵ_{dcrev} is the dc reversal error. If moderately high performance amplifiers are used for A_1 and A_4, which have a maximum input offset voltage of 1 mV each, and the resistors used in the absolute-value circuit are precision resistors with a tolerance of ± 0.1 percent, the resulting maximum error at the output is ± 4 mV $\pm 4 \times 10^{-3} E_i$. With ac inputs this introduces a totally negligible error for maximum inputs, typically 7.07 V rms, and for 10 mV rms inputs the offsets and resistor mismatches result in an error of less than ± 1 mV at the output, which is an error of only ± 0.01 percent of full scale for a 10 V full scale.

In addition to a practical absolute-value circuit, the computing rms converter in Fig. 4.6 includes protection diodes D_5, D_6, D_7, and D_8 to prevent reverse breakdown of the logging transistor emitters, and phase compensation networks C_2/R_7 and C_3/R_8. The phase compensations were designed using the procedures in Chapter 2 to provide the best possible bandwidth in A_1; a less critical, slower phase compensation is used on A_3 since it has only direct currents. Also, diode D_9 has been included in series with the collector of Q_4 to reduce its collector-base voltage to approximately zero like that of the other logging transistors in order to obtain the best matching between all four logging transistors.

4.2.2 Low-frequency errors

If it is assumed that the gain error has been compensated as previously discussed, the computing converter's transfer function simplifies to

$$E_o^2(s) = \frac{E_i^2(s)}{1 + s\tau} \qquad (4\text{-}5)$$

where τ is the transfer-function time constant, $R_2 C_1$ for the direct conversion and $R_2 C_1/2$ for the implicit conversion. For sinusoidal inputs the input squared has two components, according to Eq. (4-3): a dc component E_{irms}^2 and an ac component $E_{irms}^2 \cos 2\omega t$. Solving Eq. (4-5) for each component and then adding the components by superposition yields the magnitude of the steady-state output. The solution for the dc component is simply $e_o = E_{irms}$, and the solution for the ac component can be found by letting $s = j2\omega$ and then determining the magnitude of the transfer function:

$$\frac{|E_o^2(j2\omega)|}{|E_i^2(j2\omega)|} = \sqrt{\frac{1}{1 + 4\omega^2 \tau^2}}$$

Now the total output as a function of the input frequency is

$$e_o(t) = E_{irms} \sqrt{1 + \frac{\cos 2\omega t}{\sqrt{1 + 4\omega^2 \tau^2}}} \qquad (4\text{-}6)$$

For very low frequency inputs Eq. (4–6) simplifies to

$$e_o(t) = E_{irms} \sqrt{1 + \cos 2\omega t} \qquad \omega \ll \tau$$

and by making use of the identity

$$\cos \frac{x}{2} = \pm \sqrt{\frac{1 + \cos x}{2}}$$

and realizing that

$$e_o(t) = |\sqrt{e_o^2(t)}|$$

we obtain the expected very low frequency output

$$e_o(t) = \sqrt{2} E_{irms} |\cos \omega t| \qquad \omega \ll \frac{1}{\tau} \qquad (4\text{–}7)$$

As the frequency of the input is increased, the ripple at the output becomes smaller. If we assume the frequency of the input is greater than the inverse of the transfer-function time constant, Eq. (4–6) can be simplified by expanding it into a power series approximation. The approximation

$$\sqrt{1 + x} \doteq 1 + \frac{x}{2} - \frac{x^2}{8}$$

is sufficiently accurate for $\omega > 1/\tau$, and this allows Eq. (4–6) to be rewritten as

$$e_o(t) = E_{irms} \left[1 - \frac{1}{16(1 + 4\omega^2\tau^2)} + \frac{\cos 2\omega t}{2\sqrt{1 + 4\omega^2\tau^2}} - \frac{\cos 4\omega t}{16(1 + 4\omega^2\tau^2)} \right] \qquad \omega > \frac{1}{\tau} \qquad (4\text{–}8)$$

From this it is seen that there is both a dc error and an ac error in the output. The ac error in the output voltage, the output ripple e_{ripple}, is essentially

$$e_{\text{ripple}} = \frac{E_{irms} \cos 2\omega t}{2 \sqrt{1 + 4\omega^2\tau^2}} \qquad \omega > \frac{1}{\tau} \qquad (4\text{–}9)$$

since the fourth-harmonic ripple term is much less than the second-harmonic term with $\omega > 1/\tau$. The ripple error in the output can be reduced in many applications with a simple low-pass filter, but an error in the output dc level ϵ_{odc} will still exist, and from Eq. (4–8) it is

$$\epsilon_{\text{odc}} = \frac{E_{irms}}{16(1 + 4\omega^2\tau^2)} \qquad \omega > \frac{1}{\tau} \qquad (4\text{–}10)$$

The magnitudes of both the instantaneous error at the output due to the ac component of error in Eq. (4–8) and the error in the dc level at the

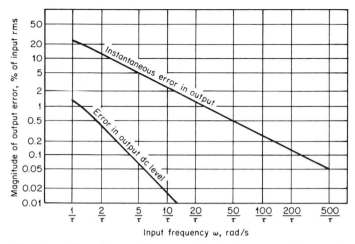

Fig. 4.7 Insufficient low-pass filtering results in both an error in the output dc level and an instantaneous error due to the ripple at the output.

output are plotted in Fig. 4.7 as a function of the input signal frequency. As is readily apparent, the dc level at the output becomes accurate much faster with increasing frequency than does the instantaneous error. Additional low-pass filtering at the output can be used to reduce the amplitude of the ripple if there is sufficient filtering inside the feedback loop to provide the required accuracy in the dc level. The low-frequency cutoff f_L as a function of the accuracy required can be determined from Eq. (4-10), and for small error it is

$$f_L = \frac{1}{16\pi\tau \sqrt{K_r}} \qquad (4\text{-}11)$$

where K_r is the fractional error ϵ_{odc}/E_{irms}; for example, $K_r = 0.01$ for 1 percent error in the output. Inserting the appropriate value of the time constant into Eq. (4-11) yields the low-frequency cutoff for both the direct and the implicit rms converter,

$$f_L = \frac{1}{16\pi R_2 C_1 \sqrt{K_r}} \qquad \text{direct rms conversion}$$

and

$$f_L = \frac{1}{8\pi R_2 C_1 \sqrt{K_r}} \qquad \text{implicit rms conversion}$$

4.2.3 Step response and settling time In order to obtain the best low-frequency performance, the transfer-function time constant is made large compared to the period of the input. This results in the desired reduction in the ripple at the output, thereby providing the desired accuracy in the output dc level. However, making the transfer-function time constant large necessarily results in a slow response at the output to a step change in the input rms level. The step response of the rms converter can be most easily analyzed by making the input a dc level which steps from an initial value E_{i1} to a final value E_{i2}. The results then apply for any input waveform, provided the lowest frequency does not cause an error in the dc level at the output which is greater than the desired setting error. The input squared is

$$e_i^2(t) = E_{i1}^2 + (E_{i2}^2 - E_{i1}^2)u(t)$$

where $u(t)$ is the unit step function. The output squared is its initial value E_{i1}^2 plus the circuit's response to the input step, $(E_{i2}^2 - E_{i1}^2)u(t)$. The total output squared is then

$$E_o^2(s) = \frac{E_{i1}^2}{s} + \frac{1}{s(1+s\tau)}(E_{i2}^2 - E_{i1}^2)$$

and the inverse Laplace transform of this is

$$e_o^2(t) = E_{i1}^2 + (1 - e^{-t/\tau})(E_{i2}^2 - E_{i1}^2)$$

Now the step response can be written

$$e_o(t) = \sqrt{E_{i1}^2\, e^{-t/\tau} + E_{i2}^2\, (1 - e^{-t/\tau})} \qquad (4\text{-}12)$$

From the step response equation the settling time t_S as a function of the settling error η can be determined:

$$\eta = \frac{e_o(\infty) - e_o(t_S)}{e_o(\infty) - e_o(0)}$$

and

$$t_S = \tau \ln \frac{E_{i1}^2 - E_{i2}^2}{[E_{i2} - (E_{i2} - E_{i1})\eta]^2 - E_{i2}^2} \qquad (4\text{-}13)$$

This expression shows the settling time to be a function of the step amplitude and direction. If the step is from a low initial value to a high one, Eq. (4-13) simplifies to

$$t_{S+} = \tau \ln \frac{1}{2\eta} \qquad E_{i2} \gg E_{i1} \text{ and } \eta \ll 1 \qquad (4\text{-}14)$$

for small settling error. The settling time for a decreasing step t_{S-}, also from Eq. (4-13), is

$$t_{S-} = 2\tau \ln \frac{1}{\eta} \qquad E_{i2} \ll E_{i1} \text{ and } \eta \ll 1 \qquad (4\text{-}15)$$

Examination of Eqs. (4–14) and (4–15) shows the settling time for decreasing step inputs to be more than double that for increasing step inputs.

If it is assumed that the maximum error in the output due to low-frequency sine-wave inputs is the same as the allowable settling error, the settling time can be expressed as a function of the accuracy required and the low-frequency cutoff. Combining Eqs. (4–11), (4–14), and (4–15) yields

$$t_{S+} = \frac{1}{16\pi f_L \sqrt{\eta}} \ln \frac{1}{2\eta}$$

and

$$t_{S-} = \frac{1}{8\pi f_L \sqrt{\eta}} \ln \frac{1}{\eta}$$

As an example, if 0.5 percent accuracy is required and the lowest-frequency sine wave to be converted is 20 Hz, the +20 dB settling time is about 65 ms, and about 150 ms is required for a −20 dB step. If the accuracy required were improved to 0.1 percent, the settling time for a +20 dB step would increase to about 195 ms, and for a −20 dB step the settling time required would be 435 ms.

4.2.4 Exact rms conversion over a time interval t_o

Early in this chapter the definition of the rms value was given as

$$E_{irms} = \sqrt{\frac{1}{t_o} \int_0^{t_o} e_i^2(t)\, dt}$$

and a practical, but subtle, approximation was made that the expression under the radical is the average value of $e_i^2(t)$. This approximation provides a continuous measure of the rms value and is most appropriate for periodic waveforms. For inputs with frequencies between direct current and the cutoff frequency of the averaging low-pass filter, there will be insufficient averaging, and errors in the output will exist, as discussed in the previous sections. So in some applications, this averaging approximation is inadequate, and the exact rms value over a fixed interval of time is required.

The computing rms converters previously discussed can easily be made to provide the exact rms value by making the low-pass filter an integrator. This is accomplished by making R_2 infinite in the rms converters shown in Figs. 4.4, 4.5, and 4.6. For the implicit rms converter previously analyzed, if R_2 is infinite the differential equation becomes

$$\frac{de_o^2}{dt} = \frac{I_{S2} I_{S4}\, \alpha_3}{I_{S1} I_{S3}\, \alpha_1} \frac{R_3}{R_1^2} \frac{2 e_i^2}{C_1}$$

which has a solution

$$e_o^2(t) = \frac{I_{S2}I_{S4}}{I_{S1}I_{S3}} \frac{\alpha_3}{\alpha_1} \frac{R_3}{R_1} \frac{2}{R_1 C_1} \int_0^{t_o} e_i^2(t)\, dt + V_{C1}(0)$$

Based on the same reasoning previously discussed, R_3 should be adjusted to force

$$\frac{I_{S2}I_{S4}}{I_{S1}I_{S3}} \frac{\alpha_3}{\alpha_1} \frac{R_3}{R_1} = 1$$

This is a calibration of the conversion gain; it is probably done most easily by applying a dc input voltage of known value and adjusting for the correct output at the end of the time interval $t_o = R_1 C_1/2$. The integrating capacitor C_1 should be shorted until $t = 0$ to force the initial condition $V_{C1}(0) = 0$. The output voltage is then

$$e_o(t) = \sqrt{\frac{2}{R_1 C_1} \int_0^{t_o} e_i^2(t)\, dt}$$

and at the time $t_o = R_1 C_1/2$, the output voltage is the rms value of the input voltage over that interval of time. The error which will result if the output is measured at a time other than $R_1 C_1/2$ is also a gain error, and if a repeatable time interval is available, its error from the desired value of $R_1 C_1/2$ can be calibrated out by adjusting R_3 as previously mentioned.

The output can be made to hold on the rms value of the input over the interval of time $t_o = R_1 C_1/2$ by shorting the collector of Q_4 in Fig. 4.6 to ground at that time. Diode D_9 will have zero voltage bias and its current will therefore be zero, and the output of amplifier A_3 will hold on the rms value of the input. The input bias current of A_3 will subtract charge from, or add charge to, the integrating capacitor, causing the output to either droop or rise. The error some time t_m after t_o due to the input bias current of the amplifier is

$$\epsilon_m = \frac{I_B t_m}{C_1}$$

where I_B is the input bias current of A_3. As an example, consider the case of a time interval $t_o = 100$ s with $R_1 = 20$ kΩ as shown in Fig. 4.6; C_1 is then 10,000 μF. An error voltage ϵ_m of 10 mV (0.1 percent of full scale for the typical 10 V full scale) will result at the end of 100 s due to an input bias current of 1 μA.

The capacitance value of 10,000 μF is probably impractical, and if long integrating times are necessary, the impedance level of the rms converter should be increased. As a result of this, it may be necessary to use FET input operational amplifiers to prevent the input bias current of the amplifiers from producing significant errors. The FET buffer shown in Fig. 2.11b and discussed in Chapter 2, Sec. 2.2.4, may be useful. If the input bias currents can be reduced to 10 pA, the impedance level can be increased such that R_1 is 2 MΩ with little or no error due to the bias currents, and now C_1 for the same integrating interval $t_o = 100$ s is only 100 μF.

4.2.5 High-frequency limitations

There are two high-frequency limitations: the bandwidth of the absolute-value circuit and the bandwidth of the input logging circuit. Because of slew-rate limitations the bandwidth of the absolute-value circuit is a function of the signal amplitude. With input signal amplitudes which are large compared with the sum of the forward drop of the rectifying diodes, D_1 and D_2 in Fig. 4.6, the errors which result because of the time required to reverse the diode states are small. However, when the amplitude of the input is small and has a high frequency, the time required to reverse the state of the rectifying diodes can produce significant error. If the input is moving to a positive level from a negative one, the state of the rectifying diodes should reverse; however, because of the finite bandwidth and slew rate the diode states do not reverse instantaneously, and during the transition time a signal current of the wrong polarity is passed. For the opposite direction of signal transition, there will be a period of time when no signal is passed, even though one should be. The amplitude of the error signal is a function of the rectifying diodes' forward drop and the amplifier's slew rate, and the resulting error is more significant with low-amplitude input signals. With large inputs the transition time to reverse the diode states may not produce significant errors, but the finite slew rate will limit the large-signal bandwidth in the conventional way, and the commonly specified full-power bandwidth will pose a high-frequency limit.

Even if the absolute-value circuit had infinite bandwidth, the rms converter's high-frequency performance would be limited by the bandwidth of the input logging circuit. The bandwidth characteristics of the logging circuit were discussed in Chapter 2, but briefly, the bandwidth is a function of the signal level, since the effective phase compensation is increased with low levels because the logging transistor's dynamic impedance is an inverse function of its collector current, $r_e = kT/qI_C$. This effect is compounded by the rectification of the input; the current in the logging transistors for even sine-wave input signals is complex, having a dc level equal to the average absolute-value of the input plus an infinite series of ac terms, with the lowest component frequency being twice the input frequency. The dynamic impedance of the logging transistors is approximately equal to $R_1 kT/qE_{iave}$, where E_{iave} is the average absolute value discussed in Sec. 4.1; this is then easily related to the rms value of the input with the sine-wave form factor, or

$$r_e \doteq \frac{1.11 R_1 kT}{qE_{irms}}$$

The small-signal −3 dB bandwidth can be found with Eq. (2-2) from Chapter 2 if the equation is modified to include two logging transistors in series:

$$f_{-3\text{ dB}} = \frac{1}{2\pi C_2(R_7 + 2.22 R_1 kT/qE_{\text{irms}})}$$

For the circuit values shown in Fig. 4.6, the −3 dB bandwidth for a full-scale input of 7.07 V rms is about 270 kHz, while it is only 185 kHz for a 70.7 mV rms input. Also, it must be recognized that this is the sine-wave −3 dB bandwidth of the logging circuit, while the actual current is an infinite series with the lowest ac component frequency being twice the frequency of the input to the absolute-value circuit.

The overall bandwidth obtained is further complicated by the fact that if the absolute-value circuit had infinite bandwidth, the limited bandwidth of the logging circuit could produce a maximum error of only about 11 percent. There is a maximum error due to limited bandwidth in the logging circuit because the dc component of current differs from the total rms value only by the sine-wave form factor of 1.11072. The net result is that the implicit rms converter's high-frequency limit is primarily due to the absolute-value circuit, which has a complex frequency response related to the amplifier's bandwidth and slew rate and the signal amplitude. The bandwidth of the absolute-value circuit is inherently inversely related to signal level, but is also limited for large signals by the amplifier's full-power bandwidth. The inverse relationship of bandwidth to signal level is further enhanced by the logging circuit's bandwidth being inversely related to the signal level. The sine-wave bandwidth typically specified for modular implicit rms converters is 0.1 percent flatness to 10 kHz, 1 percent to 80 kHz, and 3 dB to 450 kHz, for signal levels between 10 mV rms and 7.07 V rms. These bandwidths are generally based on the limitations at either end of the signal level range, and for signals in the range of 1 to 3 V rms the bandwidth obtained may be better.

4.3 Thermal RMS Converters[3,4]

A popular technique for rms measurement is to convert the energy of the signal to be measured into heat and then develop and measure the dc signal which produces an equal amount of heat. This procedure yields the rms value of a signal, and traditionally heater resistors in conjunction with thermocouples have been used to produce accurate, wide-bandwidth rms measurements. A simple fixed-gain thermal rms converter utilizing heater resistors and thermocouples is shown in Fig. 4.8. The thermocouple is electrically isolated from the heater resistor but has tight thermal

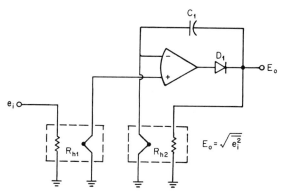

Fig. 4.8 Thermal rms conversion can be performed with matched heater resistors and thermocouples connected in a null-seeking feedback loop.

coupling to the heater. The thermally coupled heater and sensor is referred to as a *thermoelement*.

Matched pairs of thermoelements connected as in Fig. 4.8 form a null-seeking feedback loop which develops an output voltage sufficient to maintain equal thermocouple voltages. Any voltage applied to the input heater resistor produces an increase in the temperature of the input thermoelement. The increased temperature of the thermoelement produces an increase in the thermocouple voltage. This error voltage is amplified by the operational amplifier and applied to the output thermoelement's heater resistor, thus heating the output thermoelement to force a null between the input and output thermocouple voltages. For matched thermoelements a null requires that the rms value of the output voltage equal the rms value of the input voltage. With low-frequency input signals the thermal time constant of the thermoelement may be insufficient to provide a ripple-free dc output voltage, and the feedback capacitor C_1 is then used to both provide loop phase compensation and reduce the alternating current at the output. Either positive or negative polarity output will produce an increase in the output thermoelement temperature, and the diode D_1 is used to ensure that only positive outputs provide feedback.

Thermal rms converters which utilize resistor-thermocouple thermoelements have been built to provide midband accuracies as good as 0.01 percent with high-frequency limits of 100 MHz for 2 percent accuracy. However, in addition to their high cost, these thermoelements impose several performance limitations on the rms converter utilizing them. The leads of both the heater resistor and the thermocouple have a low thermal resistance which requires that the input impedance, the value of the heater resistor, be low to provide a sufficient temperature rise for accurate de-

146 Function Circuits

tection. These thermoelements also have a high thermal mass, which results in a long thermal time constant and subsequently a long output settling time. A monolithic thermoelement which uses the base-emitter junction of a bipolar transistor to sense the change in temperature of the chip due to the power dissipated by a companion diffused resistor overcomes these limitations. Low-cost monolithic and hybrid manufacturing technologies are employed, with the result that the thermal resistance of the thermoelement is high, allowing the heater resistance to be higher, and the thermal mass is very low, which allows the rms converter's output to settle faster. The remainder of this section deals primarily with the theoretical and practical considerations of thermal rms conversion utilizing a monolithic thermoelement; however, the analysis is general and is easily adapted to any form of thermoelement.

4.3.1 Monolithic thermoelement A monolithic thermoelement constructed of a silicon chip containing a single npn transistor and a diffused resistor is shown in Fig. 4.9. Figure 4.10a shows the electrical equivalent circuit of the thermoelement, including the resistor isolation and substrate junctions D_{iso} and D_{sub}, respectively. To describe the static and dynamic behavior of the thermoelement, it is convenient to model it with thermoelectrical analogs. A lumped thermoelectrical model is shown in Fig.

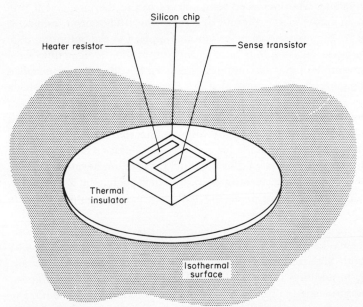

Fig. 4.9 A monolithic thermoelement is constructed of a silicon chip containing a single npn transistor and a diffused resistor thermally insulated from the ambient by an insulating disc.

RMS-to-DC Conversion   147

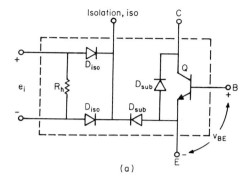

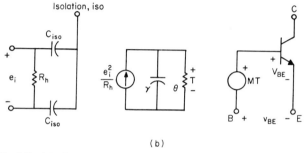

Fig. 4.10 (a) The electrical equivalent circuit of the thermoelement includes the isolation and substrate junctions. (b) A thermoelectrical model can be drawn which replaces the power dissipation by a current source, and thermal resistance and capacitance by electrical resistance and capacitance.

4.10b, where the heat flow has been replaced by a current source, heat capacity by electrical capacitance γ, and thermal resistance by electrical resistance θ. The isolation diodes D_{iso} have been replaced by their capacitance C_{iso}.

From the Ebers-Moll relationship for forward-biased junctions it can be shown that the junction has an approximately linear temperature coefficient. Equations (2–1) and (2–3) from Chapter 2 are useful, and for convenience are rewritten here.

$$v_{BE} = \frac{kT}{q} \ln \frac{I_C}{I_S} \qquad (2\text{–}1)$$

$$I_S = BT^3 e^{-qV_{go}/kT} \qquad (2\text{–}3)$$

where I_C = collector current
k = Boltzmann's constant = $8.62\,(10)^{-5}\ eV/°K$
q = charge of electron = 1 eV
T = absolute temperature
I_S = saturation current

B = temperature-independent constant related to doping levels and junction geometry

V_{g0} = bandgap voltage at 0 K = 1.11 V

By combining Eqs. (2–1) and (2–3) the junction equation can be written in terms of the physical constants and the base-emitter voltage at a specific collector current and temperature (I_{C0} and T_0, respectively).

$$v_{BE} = \frac{kT}{q} \ln \frac{I_C}{I_{C0}} \left(\frac{T_0}{T}\right)^3 + \frac{T}{T_0}\left(V_{BE0} - V_{g0}\right) + V_{g0} \quad (4\text{--}16)$$

$V_{BE0} = v_{BE}$ at I_{C0} and T_0

Differentiating Eq. (4–16) with respect to temperature yields the junction temperature coefficient, and for constant collector current it is

$$\left.\frac{dv_{be}}{dT}\right|_{I_C=\text{constant}} = \frac{V_{BE0} - V_{g0}}{T_0} - \frac{3k}{q}\left(1 + \ln \frac{T}{T_0}\right)$$

From this expression the temperature coefficient is found to be about -2.0 mV/°C and to have a nonlinearity of less than 2 percent for temperatures between 0 and 100°C. In Fig. 4.10b the temperature dependence of the junction has been modeled by a voltage source MT, where

$$M = -\left.\frac{dv_{BE}}{dT}\right|_{I_C=\text{constant}}$$

If the resistor isolation junction capacitances C_{iso} are neglected, the transfer function for the thermoelement in the complex frequency domain s can be written as

$$V_{be}(s) = V_{BE} - \frac{M\theta}{R_h} \frac{1}{1 + s\gamma\theta} E_i^2(s) \quad (4\text{--}17)$$

where V_{BE} is the ambient base-emitter voltage with zero input signal. To facilitate analysis it is convenient to define a thermal gain A_T and a thermal time constant τ_T:

$$A_T(s) = \frac{A_{T0}}{1 + s\tau_T}$$

$$A_{T0} = \frac{M\theta}{R_h}$$

$$\tau_T = \gamma\theta$$

which simplifies Eq. (4–17) to

$$V_{be}(s) = V_{BE} - \frac{A_{T0}}{1 + s\tau_T} E_i^2(s) \quad (4\text{--}18)$$

TABLE 4.2 Dimensions and Characteristics of a Typical Monolithic Thermoelement

Parameter	Typical value
Silicon chip dimensions	24W x 24L x 6T mil
Insulator dimensions	100DIA x 2T mil
Silicon thermal conductivity	0.20 cal/s · cm · °C
Silicon thermal capacity	0.43 W · s/°C · cm³
Insulator thermal conductivity	0.00037 cal/s · cm · °C
Insulator thermal capacity	0.37 W · s/°C · cm³
Heater resistance R_h	115 Ω
Effective thermal resistance θ	575°C/W
Junction temperature sensitivity M	2 mV/°C
Thermal gain, A_{T0}	10 mV/V²
Effective thermal capacitance, γ	113 μW · s/°C
Thermal time constant, τ_T	65 ms

At this point it should be noted that the thermal gain A_{T0} is nonlinear because of the slight nonlinearity of the junction temperature coefficient M as discussed above, but also because of the temperature coefficients of the heater resistance R_h and the thermal resistance θ. This nonlinearity has little effect on the dynamic behavior of the thermoelement; however, it can produce some errors in the dc response.

The typical characteristics measured from a monolithic thermoelement are shown in Table 4.2. Note that if the thermal resistance is calculated from the chip and insulator dimensions, a value of approximately 880°C/W is obtained. The discrepancy between the calculated and measured thermal resistance is probably due to heat loss through the wirebonds. Also, the calculated thermal capacitance γ is in good agreement with the measured value if the volume of the insulator heated is assumed to be the area of the chip times only one-half the insulator thickness, since the full mass of the insulator is not heated to the temperature of the chip.

4.3.2 Midband and dc transfer function A rms-to-dc converter can be built using a pair of thermoelements, and a suitable circuit is shown in Fig. 4.11. The power due to the input signal e_i dissipated by the heater resistor R_{h1} heats the input thermoelement, producing a change in the base-emitter voltage of Q_1. This generates an error voltage which is amplified by Q_1, Q_2, and operational amplifier A_1. The amplified error voltage is applied to R_{h2} and heats the output thermoelement, thus tending to bring the circuit into equilibrium. When the circuit is in equilibrium, $E_{orms} = E_{irms}$.

For high-frequency input signals the thermal time constant of the input thermoelement and the finite bandwidth of operational amplifier A_1 act to average the input power, so that E_o is a dc voltage. For slowly varying input signals, the temperature of Q_1 will track the instantaneous value of e_i^2, and E_o would tend to follow e_i. Negative ac feedback from the output

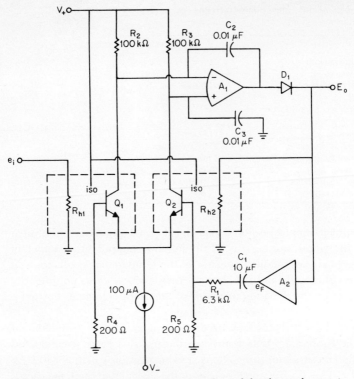

Fig. 4.11 Thermal rms converter using a pair of monolithic thermoelements in a null-seeking feedback loop. (*Koerner, U.S. patent 3,668,428, 1972.*)

is provided by amplifier A_2, C_1, and R_1; this forces the output to be a dc voltage for even very low frequency inputs.

Capacitors C_2 and C_3 are included to provide phase compensation for the composite operational amplifier formed by Q_1, Q_2, R_2, R_3, and A_1. For the circuit values shown there is sufficient phase margin that 100 percent ac feedback ($R_1 = 0$, $A_2 = 1$) will not produce oscillations. Provided C_2 and C_3 do not become so large that they reduce the open-loop gain to unity before about 10 kHz, they do not affect the closed-loop response, except that they provide phase margin.

Neglecting the effects of the sense transistor base currents in the base resistors R_4 and R_5, the base-emitter voltages of the sense transistors must be related as

$$V_{\text{be1}}(s) = V_{\text{be2}}(s) - H(s)E_o(s) \tag{4-19}$$

where $H(s)$ is the ac feedback transfer function,

$$H(s) = \frac{sR_5C_1A_2}{1 + sC_1(R_5 + R_1)} \tag{4-20}$$

If we make

$$R_1 = \frac{\tau_{T2}}{C_1} - R_5$$

and define a filter time constant τ_F

$$\tau_F = \frac{R_5 C_1 A_2}{A_{T02} E_o(s)} \qquad (4\text{–}21)$$

combining the thermoelement transfer function, Eq. (4–18), with Eqs. (4–19) and (4–20) yields

$$E_o{}^2(s) = \frac{A_{T01}(1 + s\tau_{T2})}{A_{T02}(1 + s\tau_{T1})} \frac{E_i{}^2(s)}{1 + s\tau_F} + \frac{V_{BE2} - V_{BE1}}{A_{T02}} \qquad (4\text{–}22)$$

Note that the frequency dependence of Eq. (4–22) is nonlinear, since the transfer-function time constant, Eq. (4–21), is inversely proportional to the output voltage. This characteristic will be dealt with in the following sections on step response and low-frequency performance.

For sinusoidal inputs,

$$e_i(t) = \sqrt{2} E_{irms} \cos \omega t$$

and

$$e_i{}^2(t) = E_{irms}{}^2(1 + \cos 2\omega t) \qquad (4\text{–}23)$$

If the frequency of the input is much greater than the inverse of the filter time constant τ_F and the thermal time constants τ_{T1} and τ_{T2}, or if the input is a dc signal, the steady-state transfer function simplifies to

$$E_o = \sqrt{\frac{A_{T01}}{A_{T02}} E_{irms}{}^2 + \frac{V_{BE2} - V_{BE1}}{A_{T02}}} \qquad (4\text{–}24)$$

From Eq. (4–24) it can be seen that there are two error sources in the midband rms-to-dc conversion: (1) mismatch in the thermal gains, which produces a gain error that can be compensated by a scale factor adjustment in the dc measuring circuit applied to the output, and (2) the ambient mismatch in the base-emitter voltages of the sense transistors Q_1 and Q_2, which produces a nonlinearity in the rms-to-dc conversion. Because of the high gain of A_1, the voltages across the collector load resistors R_2 and R_3 are forced to be essentially equal, which causes the ambient mismatch in base-emitter voltages to be

$$V_{BE2} - V_{BE1} = \frac{kT_a}{q} \ln \left(\frac{I_{S1}}{I_{S2}} \frac{R_2}{R_3} \right)$$

where T_a is the ambient temperature in degrees Kelvin. It is apparent from this expression that the error in the rms-to-dc conversion due to offset voltage between the sense transistors is easily nulled by adjusting the ratio of the collector load resistors R_2 and R_3.

The output stability as a function of the thermal gain drift and the offset voltage drift can be derived from the midband and dc transfer function, Eq. (4–24), and an approximate expression is

$$\Delta E_o \doteq \frac{\Delta(A_{T01}/A_{T02})}{2} E_{irms} + \frac{\Delta(V_{BE2} - V_{BE1})}{2A_{T02}E_{irms}} \qquad (4\text{--}25)$$

Over an input signal range of 0.1 to 3.0 V rms, the output stability versus temperature for the rms-to-dc converter is typically $(50E_{irms} + 50/E_{irms})$ μV/°C, where E_{irms} is in volts. The lower limit on the input range of 0.1 V rms is imposed because of the $1/E_{irms}$ error component in Eq. (4–25). The upper limit of 3.0 V rms is to prevent damaging the thermoelement by overheating. Typical computing rms-to-dc converters discussed in previous sections of this chapter do not have these limitations and remain reasonably accurate over a wider dynamic range than the thermal converter.

If the offset voltage, $V_{BE2} - V_{BE1}$, is compensated by adjusting R_2 and R_3 as discussed above, Eq. (4–24) indicates the only remaining midband error to be a gain error due to mismatch between A_{T01} and A_{T02}. However, these thermal gains are slightly nonlinear as a result of the temperature coefficient of both the heater resistor and the thermal resistance, and also as a result of the nonlinearity of the junction temperature coefficient, as discussed. The nonlinearity of the thermal gains would have no effect, to the extent that they are matched and therefore cancel each other. The inherently low nonlinearity of the thermal gain and good repeatability of the monolithic thermoelements allows the nonlinearity of the rms conversion to typically be reduced to ±0.025 percent of full scale by adjusting the load resistors R_2 and R_3, with $E_{irms} = 0.1$ V rms.

A further source of midband and dc error is the voltage nonlinearity of the diffused resistor. Voltage nonlinearity in a diffused resistor results from the diode junction isolation; the reverse bias on this junction varies with the input, causing the depletion into the diffused resistor to vary, thereby modulating the value of the resistor. The voltage nonlinearity of the heater resistor causes the rms converter to have a dc reversal error of 0.04 percent. The dc reversal error is defined as the change in the output due to reversing the polarity of a dc input. If the unit is calibrated with a sine-wave input, the dc reversal error only causes a ±0.02 percent error on a dc input signal component.

Voltage nonlinearity of the heater resistor also produces errors with high crest factor. High-crest-factor inputs also cause an additional form of heater resistor voltage nonlinearity: The low thermal mass of the monolithic thermoelement allows the temperature of the input heater resistor to track the input voltage squared for even relatively high-frequency inputs. This large variation of the heater resistor temperature induces a large variation in the value of the heater resistor because of the typically high temperature coefficient of a diffused resistor (1000 ppm/°C). The thermally induced voltage nonlinearity and the depletion spread variations with voltage combine to produce approximately a 1.5 percent error for an input pulse train with a 1 ms period and a 5:1 crest factor. When the repetition rate of the input is increased to 10 kHz, the thermally induced voltage nonlinearity is effectively eliminated, and the conversion accuracy improves to 0.5 percent.

The voltage nonlinearity of a diffused resistor due to both depletion spread into the resistor and large temperature drift of the resistance with high-crest-factor inputs could be effectively eliminated by using a compatible thin-film heater resistor. By using a low-drift (± 50 ppm/°C) thin-film resistor with glass isolation, the dc reversal error should be reduced to less than 0.01 percent, and the high-crest-factor performance should be improved to 0.3 percent with even relatively low-frequency high-crest-factor pulse trains.

4.3.3 Low-frequency errors

As mentioned above, the frequency dependence of the rms-to-dc converter shown in Fig. 4.11 may be nonlinear, since the transfer-function time constant is inversely proportional to the output voltage if A_2 is linear. As will be shown later, this nonlinear frequency dependence produces an unsymmetrical step response beyond that discussed for the computing rms converter, and a low-frequency cutoff for accurate rms conversion which is proportional to the rms value of the input. These characteristics are undesirable if fast response is necessary; a nonlinear ac feedback should be employed. If A_2 is made a square-law amplifier,

$$A_2 e_o(t) = \rho e_o^2(t) \qquad (4\text{-}26)$$

the time constant will be independent of the output; combining Eqs. (4-21) and (4-26), the filter time constant for square-law A_2, τ_{FS}, is

$$\tau_{FS} = \frac{R_5 C_1 \rho}{A_{T02}} \qquad (4\text{-}27)$$

Assuming that the mismatch between thermal gains and base-emitter voltages of the two thermoelements has been compensated, the transfer function for the thermal rms converter simplifies to

$$E_o^2(s) = \frac{1 + s\tau_{T2}}{1 + s\tau_{T1}} \frac{E_i^2(s)}{1 + s\tau_{FS}} \qquad (4\text{-}28)$$

with square-law ac feedback. Note that with the exception of the pole-zero pair due to the thermal time constants, which generally cancel since τ_{T1} usually is approximately equal to τ_{T2}, this expression is identical to the one obtained for the computing rms converter [Eq. (4-5)]. As with the computing rms converter, the magnitude of the steady-state output with sine-wave inputs is most easily determined by solving Eq. (4-28) for the dc component of the input squared, E_{irms}^2, and also for the ac component, $E_{irms}^2 \cos 2\omega t$, and then adding them by superposition to obtain the total output as a function of frequency. Performing these manipulations yields the steady-state output with square-law ac feedback:

$$e_o(t) = E_{irms} \sqrt{1 + \left[\frac{1 + 4\omega^2 \tau_{T2}^2}{(1 + 4\omega^2 \tau_{T1}^2)(1 + 4\omega^2 \tau_{FS}^2)} \right]^{1/2} \cos 2\omega t} \qquad (4\text{-}29)$$

For very low frequency inputs this yields an expression for the output which is identical to that for the computing rms converter [see development of Eq. (4-7) in Sec. 4.2.2]:

$$e_o(t) = \sqrt{2} E_{irms} |\cos \omega t| \qquad \omega \ll \frac{1}{\tau_{FS}}$$

As the frequency of the input is increased, the ripple at the output becomes smaller. Using the same power series approximation on Eq. (4-29) for small ripple as was done for the computing converter yields the ac error in the output voltage. The output ripple e_{ripple} is

$$e_{\text{ripple}} = \frac{\tau_{T2}}{\tau_{T1}} \frac{E_{irms} \cos 2\omega t}{4\omega \tau_{FS}} \qquad \omega > \frac{1}{\tau_{FS}}, \frac{1}{\tau_{T1}}, \frac{1}{\tau_{T2}} \qquad (4\text{-}30)$$

and the dc error in the output voltage ϵ_{odc} for small error is

$$\epsilon_{\text{odc}} = \frac{\tau_{T2}}{\tau_{T1}} \frac{E_{irms}}{64\omega^2 \tau_{FS}^2} \qquad \omega > \frac{1}{\tau_{FS}}, \frac{1}{\tau_{T1}}, \frac{1}{\tau_{T2}} \qquad (4\text{-}31)$$

For both these expressions, Eqs. (4-30) and (4-31), a further approximation that $4\omega t \gg 1$ was implemented to simplify the equations; however, the results are identical to those for the computing converter except for the τ_{T2}/τ_{T1} factor. Because of the repeatability of the monolithic thermoelement, $\tau_{T1} \doteq \tau_{T2}$, and therefore τ_{T2}/τ_{T1} will be considered to be unity in the following analysis.

The analysis leading to Eq. (4-29) is invalid for filter time constants which are a function of the output, as is the case with linear A_2. An approximate solution for linear A_2 can be obtained by relating the output of the linear amplifier to that of the square-law amplifier:

$$\begin{aligned} e_F &= A_2 e_o \\ e_{Fac} &= A_2 e_{\text{ripple}} \end{aligned} \qquad \text{linear } A_2 \qquad (4\text{-}32)$$

and

$$e_F = \rho e_o^2.$$
$$e_{Fac} \doteq 2\rho E_{odc} e_{ripple} \quad \text{square-law } A_2 \quad (4\text{-}33)$$

From Eqs. (4-32) and (4-33) it is apparent that the square-law amplifier's gain is effectively $A_2/2E_{odc}$. Combining this result with Eqs. (4-21) and (4-30) yields the peak magnitude of the output ripple,

$$|e_{ripple}| = \frac{A_{T02} E_{odc} E_{irms}}{4\pi f R_5 C_1 A_2} \quad \text{linear } A_2 \quad (4\text{-}34\text{a})$$

$$|e_{ripple}| = \frac{A_{T02} E_{irms}}{8\pi f R_5 C_1 \rho} \quad \text{square-law } A_2 \quad (4\text{-}34\text{b})$$

where f is the input frequency in hertz and is greater than $1/\tau_F$.

The error in the dc voltage at the output can also be found with the approximate relationship between the linear and square-law ac feedback gains; combining it with Eqs. (4-21) and (4-31) yields

$$\epsilon_{odc} = \frac{A_{T02}^2 E_{odc}^2 E_{irms}^2}{64\pi^2 f^2 R_5^2 C_1^2 A_2^2} \quad \text{linear } A_2 \quad (4\text{-}35\text{a})$$

$$\epsilon_{odc} = \frac{A_{T02}^2 E_{irms}^2}{256\pi^2 f^2 R_5^2 C_1^2 \rho^2} \quad \text{square-law } A_2 \quad (4\text{-}35\text{b})$$

The low-frequency cutoff f_L as a function of the accuracy in the dc level at the output can be determined from Eq. (4-35):

$$f_L = \frac{A_{T02} E_{odc}}{8\pi R_5 C_1 A_2 \sqrt{K_r}} \quad \text{linear } A_2 \quad (4\text{-}36\text{a})$$

$$f_L = \frac{A_{T02}}{16\pi R_5 C_1 \rho \sqrt{K_r}} \quad \text{square-law } A_2 \quad (4\text{-}36\text{b})$$

where K_r is the fractional error ϵ_{odc}/E_{irms}. As expected, the low-frequency cutoff for accurate rms-to-dc conversion is roughly proportional to the rms value of the input with linear ac feedback, where f_L is independent of the signal level with square-law ac feedback.

In Fig. 4.12a the peak value of the output ripple is plotted versus the rms value of the input for an input frequency of 20 Hz. The circuit values are those shown in Fig. 4.11, with the ac feedback amplifier being either a short ($A_2 = 1$) or square-law with $e_F = 0.25 e_o^2$ ($\rho = 0.25$). As expected, the measured output ripple with linear ac feedback is proportional to the square of the rms value of the input, where it is linearly proportional to the rms value of the input with square-law ac feedback. The square-law

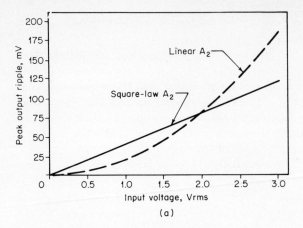

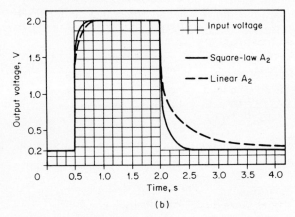

Fig. 4.12 (a) Output ripple versus input level for a 20 Hz sine-wave input with both linear and square-law ac feedback. (b) Step response of rms converter with linear and square-law ac feedback. ($R_1 = 6.3$ kΩ, $R_5 = 200$ Ω, $C_1 = 10$ μF, and $A_2E_o = E_o$ or $A_2E_o = 0.25E_o^2$.)

ac feedback does, therefore, provide a low-frequency cutoff which is independent of the signal level, as it is with the computing rms converters.

4.3.4 Step response and settling time Both modes of ac feedback discussed for thermal rms converters, linear and square-law, are equally capable of reducing the ripple at the output to any degree required. Since the linear ac feedback is so much simpler to implement than the square-law ac feedback, it may not be clear why we even consider square-law ac feedback. As was briefly stated before, the transfer-function time constant is an inverse function of the output voltage with linear ac feedback, and this results in a very unsymmetrical step response. As will be shown later in this section, the step response for increasing rms level is only slightly affected by the mode of ac feedback; however, if the step is from a high rms level to a low one, the filter time constant is continuously lengthening, and the negative step response may be very long. If the lowest-frequency input to be measured is relatively high, about 400 Hz or greater, or the negative step response is unimportant, simple linear ac feedback with $A_2 = 1$, or simply a short, will probably be adequate. As will be discussed later, second-order effects due to mismatch between the thermal time constants and thermal gradients between the monolithic thermoelements will limit the -20 dB step response to requiring about 0.6 s for settling to 0.1 percent, even if square-law ac feedback is used. Therefore, if in Fig. 4.11 we make $R_1 = 130$ kΩ, $R_5 = 200$ Ω, $C_1 = 0.5$ μF, and A_2 simply a short, from Eq. (4-36a) the low-frequency cutoff with 3 V rms input sine waves will be about 400 Hz for 0.1 percent accuracy, and the settling time for a -20 dB step will be close to the minimum obtainable: 0.6 s to 0.1 percent. However, if a lower-frequency input must be converted and the settling time is important, square-law ac feedback should be employed to obtain a better settling time and low-frequency cutoff product.

As with the computing rms converter, the step response is most easily determined by considering the case where the input is a dc level which steps from an initial value E_{i1} to a final value E_{i2}. Again, this result then applies for any waveform, provided that the lowest frequency does not produce an error in the output dc level which is greater than the settling error. The input squared is

$$e_i^2(t) = E_{i1}^2 + (E_{i2}^2 - E_{i1}^2)u(t)$$

where $u(t)$ is the unit step function. The output squared is the initial value E_{i1}^2 plus the circuit's response to the input step $(E_{i2}^2 - E_{i1}^2)u(t)$, as it was for the computing rms converter. The total output squared is then

$$E_o^2(s) = \frac{E_{i1}^2}{s} + \frac{1 + s\tau_{T2}}{1 + s\tau_{T1}} \frac{(E_{i2}^2 - E_{i1}^2)}{(1 + s\tau_{FS})s}$$

and the inverse Laplace transform of this is

$$e_o^2(t) = E_{i1}^2 + \left(1 - \frac{\tau_{FS} - \tau_{T2}}{\tau_{FS} - \tau_{T1}} e^{-t/\tau_{FS}} + \frac{\tau_{T1} - \tau_{T2}}{\tau_{FS} - \tau_{T1}} e^{-t/\tau_{T1}}\right)(E_{i2}^2 - E_{i1}^2)$$

Now the step response for the thermal rms converter with square-law ac feedback can be written:

$$e_o(t) = \left[\left(\frac{\tau_{FS} - \tau_{T2}}{\tau_{FS} - \tau_{T1}} e^{-t/\tau_{FS}} - \frac{\tau_{T1} - \tau_{T2}}{\tau_{FS} - \tau_{T1}} e^{-t/\tau_{T1}}\right) E_{i1}^2 + \left(1 - \frac{\tau_{FS} - \tau_{T2}}{\tau_{FS} - \tau_{T1}} e^{-t/\tau_{FS}}\right.\right.$$
$$\left.\left. + \frac{\tau_{T1} - \tau_{T2}}{\tau_{FS} - \tau_{T1}} e^{-t/\tau_{T1}}\right) E_{i2}^2\right]^{1/2} \quad (4\text{-}37)$$

Examination of this reveals that if the thermal time constants are matched, the step response of the thermal rms converter with square-law ac feedback is identical to that of the computing rms converter given in Eq. (4–12). Even if the thermal time constants are not matched perfectly, Eq. (4–37) still simplifies to Eq. (4–12) if the filter time constant is large compared to the mismatch:

$$e_o(t) \doteq \sqrt{E_{i1}^2 e^{-t/\tau_{FS}} + E_{i2}^2(1 - e^{-t/\tau_{FS}})} \qquad \tau_{FS} \gg |\tau_{T1} - \tau_{T2}|$$

The settling time is also the same as for the computing rms converter:

$$t_S \doteq \tau_{FS} \ln \frac{E_{i1}^2 - E_{i2}^2}{[E_{i2} - (E_{i2} - E_{i1})\eta]^2 - E_{i2}^2} \qquad \tau_{FS} \gg |\tau_{T1} - \tau_{T2}| \quad (4\text{--}38)$$

The analysis leading to Eq. (4–37) and subsequently to Eq. (4–38) is invalid for a transfer-function time constant which is a function of the output voltage, as it is with linear ac feedback. However, if the settling error is small, the time constant can be approximated as a constant, since the output is at essentially its final value for most of the time. By comparing the output of A_2 for both linear and square-law gain, a correction for linear A_2 can be obtained:

$$e_F = A_2 e_o$$
$$\frac{de_F}{de_o} = A_2 \qquad \text{linear } A_2 \qquad (4\text{--}39)$$

and

$$e_F = \rho e_o^2$$
$$\frac{de_F}{de_o} = 2\rho e_o \qquad \text{square-law } A_2 \qquad (4\text{--}40)$$

From Eqs. (4–39) and (4–40) it is apparent that the square-law amplifier's gain is effectively $A_2/2E_{o2}$ for most of the settling time (note the similarity to the effective value of ρ in the low-frequency cutoff analysis). Combining this result with Eqs. (4–21), (4–27), and (4–38) leads to the results for increasing steps

$$t_{S+} \simeq \frac{R_5 C_1 A_2}{2 A_{T02} E_{o2}} \ln \frac{1}{2\eta} \qquad \text{linear } A_2 \qquad (4\text{--}41a)$$

$$t_{S+} = \frac{R_5 C_1 \rho}{A_{T02}} \ln \frac{1}{2\eta} \qquad \text{square-law } A_2 \qquad (4\text{--}41b)$$

and for decreasing steps

$$t_{S-} \simeq \frac{R_5 C_1 A_2}{A_{T02} E_{o2}} \ln \frac{1}{\eta} \qquad \text{linear } A_2 \qquad (4\text{--}42a)$$

$$t_{S-} = \frac{2 R_5 C_1 \rho}{A_{T02}} \ln \frac{1}{\eta} \qquad \text{square-law } A_2 \qquad (4\text{--}42b)$$

for small settling error η and large filter time constants compared to the mismatch between the thermal time constants. The settling time for both ac feedback modes is unsymmetrical, having the same form as the settling time equation for computing rms converters. However, the settling time for the linear ac feedback case is much more unsymmetrical, since the effective filter time constant is an inverse function of the output voltage.

In Fig. 4.12b the step response for both a +20 and a −20 dB input step is shown for both linear and square-law ac feedback. For linear ac feedback A_2 was made a short ($A_2 = 1$), and for square-law ac feedback $e_F = A_2 e_0 = 0.25 e_0^2$ ($\rho = 0.25$), and the circuit values were those shown in Fig. 4.11. These values provide an effective filter time constant of $100/E_o$ ms (E_o in volts) and 50 ms for the linear and square-law cases, respectively. These time constants provide similar settling times for 2.0 V final values, according to Eq. (4–41), and as was shown in the previous section, they also provide the same output ripple for 2.0 V rms inputs. As stated earlier, the step response for linear ac feedback is very unsymmetrical, with a decreasing step of 20 dB requiring more than 20 times as long to settle as an increasing step. It is seen that the step response with square-law ac feedback is much more symmetrical, being similar to that of the computing rms converter; it does require a little more than twice as long to settle for decreasing steps as it does for increasing steps.

As mentioned at the beginning of this section, second-order effects will impose a lower limit on the settling time due to small errors. Equation (4–37) was simplified on the assumption that the filter time constant is large compared to the mismatch between the thermal time constants, and this in itself is a limit. Because of the inherently low thermal mass and the repeatability of monolithic thermoelements, the thermal time constants of the thermoelements typically cause the output to require only about 600 ms to settle to within 0.1 percent of step, for a −20 dB step input and square-law ac feedback. A further limitation on the settling time results from the fact that a high rms level can cause the package of the thermoelement to heat up, and even more significantly, thermal gradients between the two thermoelements may result. The package temperature rise and the thermal gradient between the two thermoelements have a very long thermal time constant, of the order of a minute, and produce an error in the output according to Eq. (4–25) which settles with this long time constant. This effect is most pronounced when the rms input is reduced from a high level to a low level. By packaging the two thermoelements in a common package with very close thermal proximity and good heat sinking to ambient, the error which must settle with this long thermal time constant can be reduced to less than 0.1 percent of step for a −20 dB step, and to essentially zero for a +20 dB step. Further, to obtain the optimum settling time it may be necessary to adjust the value of R_1 for each thermoelement to compensate for the variations in the thermal time constant.

4.3.5 High-frequency limitations The primary high-frequency limitations of the rms-to-dc converter in Fig. 4.11 are the distributed capacitance on the input heater resistor and capacitive coupling of high-frequency signals into the dc circuitry. Since the total distributed capacitance on the heater resistor typically is only 6 pF ($C_{iso} = 3$ pF in the model shown in Fig. 4.10b), the -3 dB input frequency would be approximately 1 GHz when driven by a 50 Ω source. A more significant problem is isolating the sensitive dc circuit from the high-frequency input, since even very small high-frequency swings in the dc circuit can cause changes in the input offset voltage of A_1, or even saturate its input stage. A complete discussion of the causes of the change in the input offset voltage of A_1 is beyond the scope of this chapter; however, the major cause is that parasitic capacitances in the input stage shunt the input stage current with even small high-frequency swings. A change in the input offset voltage of A_1, whether from temperature variations, time, or RF as mentioned, will cause a change in $V_{BE2} - V_{BE1}$, producing an error in the dc output as shown by Eq. (4–25). With the circuit values shown in Fig. 4.11, the change in $V_{BE2} - V_{BE1}$ is approximately one two-hundredth the change in the input offset voltage of A_1.

In order to minimize the RF at the inputs to A_1, the rms converter should be built on a ground plane with high-frequency bypass capacitors of about 1 nF as isolation from the resistor and from the collectors, bases, and emitters of the sense transistors. Proper high-frequency connectors should be used at the input, with this lead and those of the bypass capacitors kept as short as possible. To further prevent RF in the dc circuit, the ground side of the base resistors R_4 and R_5 should be connected to the ground plane at a common point, thereby preventing any ground-loop interference. Through the use of these wiring techniques the rms converter provides a ±0.05 percent bandwidth of 10 MHz and a ±2 percent bandwidth of 100 MHz.

4.3.6 Input protection for thermal rms converters All thermal converters require some form of input protection, since application of an excessive rms voltage can easily damage the converter's input heater resistor. The monolithic thermoelement described in the previous sections requires the input voltage to be limited according to the expression

$$\text{Maximum } e_i = \sqrt{\frac{125°C - T_a}{6}} \quad \text{V rms}$$

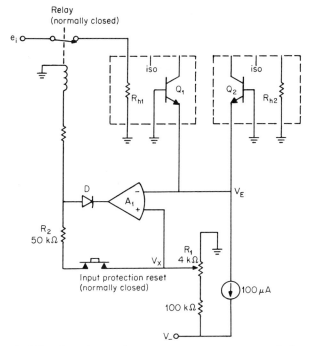

Fig. 4.13 Thermal-responding input protection circuit passes high-crest-factor inputs, while limiting the rms value of the input by opening a protection relay driven by A_1.

If only sine waves and other low-crest-factor signals are to be applied to the thermoelement, a simple diode clamp in parallel with the input may be sufficient input protection. However, if signals with high crest factor are to be measured, the input protection should be capable of limiting the rms value of the input signal while allowing low-duty-cycle signals with peak values much greater than their rms value to be measured.

The thermal-responding input protection circuit of Fig. 4.13 prevents input signals with excessive rms value from being applied to the thermo-element. With an ambient temperature T_a of 25°C and zero input voltage, R_1 should be adjusted for a voltage V_X which is 200 mV more positive than V_E. This allows input signals to be applied to the thermoelement if the thermoelements are at a temperature below 125°C. For any combination of ambient temperature and input signal that causes the thermoelement temperature to exceed about 125°C, the protection amplifier's output will go to its negative saturation level, opening the contacts of the input protection relay. Resistor R_2 provides positive feedback for latching the protection amplifier so that the relay contacts remain open until the input protection reset switch is opened.

4.4 Applying RMS Converters

Modular and integrated-circuit rms converters are available as complete building blocks, requiring only power-supply connections in addition to the input and output connections. Frequently, in addition to these connections provisions are made to allow user adjustment of the offset voltage and the gain; also, often the capacitor which sets the low-frequency cutoff is external, allowing the user to choose the optimum low-frequency cutoff and settling time for a particular application. This section deals with some considerations involved in using rms converters; it is not a section of applications, as such. Some specialized applications of rms converters, such as power and gain measurement, are discussed in Chapter 7. This section is meant to provide further information for using rms converters; such topics as reducing the output ripple, input ranging, and the requirements for measuring high-crest-factor signals are discussed.

4.4.1 Reducing the output ripple The ripple voltage at the output of both the computing and thermal rms converters can be reduced to any level required with the techniques discussed in Secs. 4.2.2 and 4.3.3. These techniques provide low-pass filtering inside the feedback loop and, therefore, ensure that the dc level at the output is an accurate measure of the rms value of the input. However, even if there is sufficient low-pass filtering inside the feedback loop to ensure that the dc level is accurate, the peak amplitude of the ripple at the output will be large for low-frequency inputs. The peak amplitude of the ripple is the error in the output if instantaneous measurements are made of the output voltage. Figure 4.7 shows that the instantaneous error in the output is much larger than the error in the dc output level, and it falls off with increasing frequency much more slowly than the error in the output dc level. If the output of the rms converter is measured with a dc voltmeter, the ripple voltage, or instantaneous error, is probably not significant, since the integrating time or dampening in the dc voltmeter will allow it to extract the dc component at the output, and the resulting error is only that in the dc level at the output.

If the output voltage is sampled by an analog-to-digital converter for processing by a digital computer or is used as a signal in a control loop, it will probably be necessary to reduce the ripple at the output to minimize the instantaneous error in the output voltage. If the low-pass filter inside the feedback loop is made to have a cutoff low enough to provide the desired accuracy not only at the lowest frequency in the dc level, but also in the instantaneous output voltage, the rms converter's response time will be made much longer: about 63 times longer for 0.1 percent accuracy.

If additional low-pass filtering is added outside the feedback loop, the instantaneous error can be reduced without lengthening the response time nearly as much as if it were added inside the feedback loop. The filtering inside the feedback loop should be sufficient to obtain the desired accuracy only in the dc level; the required filtering can be determined with Eqs. (4–11) and (4–36) for the computing and thermal rms converters, respectively. This amount of filtering inside the feedback loop will provide the

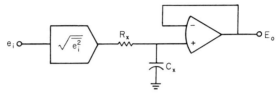

Fig. 4.14 Even a simple single-pole active filter can greatly reduce the output ripple.

essential accuracy in the rms-to-dc conversion, while permitting the minimum settling time. Figure 4.14 shows an rms converter followed by a simple one-pole active filter. The ac voltage at the input of the active filter is the output ripple voltage e_{ripple}, and from Eqs. (4–9) and (4–30) the amplitude of the ripple for input frequencies above the internal low-pass cutoff is

$$e_{\text{ripple}} = \frac{E_{\text{irms}} \cos 2\omega t}{4\omega \tau_{\text{eff}}} \qquad (4\text{--}43)$$

where τ_{eff} is the effective internal filter time constant:

$$\tau_{\text{eff}} = R_2 C_1 \qquad \text{direct rms conversion} \qquad (4\text{--}44\text{a})$$

$$\tau_{\text{eff}} = \frac{R_2 C_1}{2} \qquad \text{implicit rms conversion} \qquad (4\text{--}44\text{b})$$

$$\tau_{\text{eff}} = \frac{R_5 C_1 A_2}{2 A_{T02} E_{\text{odc}}} \qquad \text{thermal converter with linear } A_2 \qquad (4\text{--}44\text{c})$$

$$\tau_{\text{eff}} = \frac{R_5 C_1 \rho}{A_{T02}} \qquad \text{thermal converter with square-law } A_2 \qquad (4\text{--}44\text{d})$$

The ripple at the overall output e_{oac} is

$$e_{\text{oac}} = \frac{E_{\text{irms}} \cos 2\omega t}{8\omega^2 \tau_{\text{eff}} R_x C_x}$$

if $\omega > 1/R_x C_x$. (Note that the ac signal at the input of the filter has a frequency twice that of the input signal.) If $R_x C_x$ is made equal to the effective internal time constant τ_{eff}, the ripple at the output will be reduced substantially, and the settling time will only be increased about 40 percent, since as a rule of thumb

$$t_{So} \simeq \sqrt{t_{S1}^2 + t_{S2}^2 + \cdots + t_{Sn}^2}$$

where t_{So} is the overall settling time and $t_{S1}, t_{S2}, \ldots, t_{Sn}$ are the settling times of cascaded stages. The settling time of the single-pole active filter is

$$t_{Sx} = R_x C_x \ln \frac{1}{\eta}$$

where η is the settling error; for example, if settling to within 0.1 percent is required, $\eta = 0.001$.

By using more complex active filters at the output of the rms converter, the ripple voltage can be reduced even more, while the settling time is degraded only slightly. Appropriate active filters are discussed in Chapter 6.

4.4.2 Input ranging Many applications of rms converters involve signals in the range of 10 mV to 10 V, and the computing rms converters can be used directly. Frequently the more limited input range of the thermal rms converters (0.1 to 3 V rms) is also adequate. However, in many applications the signal to be converted is outside these dynamic ranges, and some input preconditioning is necessary. If the application requires only the moderate bandwidth of a computing rms converter, typically 0.1 percent to 10 kHz and 3 dB to 450 kHz, design of the input ranging circuit is quite straightforward. If the ultra-wide bandwidth of the thermal rms converter is necessary, typically as good as 2 percent to 100 MHz, the design of the input ranging circuit is more complex. The design of ultra-wide-bandwidth amplifiers is beyond the scope of this section; however, there are some significant considerations even if modular or integrated-circuit ultra-wide-bandwidth operational amplifiers are used.

A fully developed input ranging circuit for a wide-bandwidth thermal rms converter is shown in Fig. 4.15. It provides a gain of 10 for the low range of 0.2 V rms full scale and a gain of 0.01 for the high range of 200 V rms full scale. There are four ranges, allowing accurate rms-to-dc conversion of signals as low as 10 mV rms and as high as 300 V rms. The complete converter has a 50 percent over range capability as a result of 3.0 V rms full-scale capability of the thermal converter. The gain is broken into two stages to provide the maximum bandwidth. Since A_2 is always in a closed-loop gain of 10, it can be phase-compensated for this gain, and therefore it can have a better bandwidth than if it were phase-compensated for 100 percent feedback. Also, since its input is buffered by the preamplifier A_1, its feedback resistors can be kept low, alleviating the need for shunt

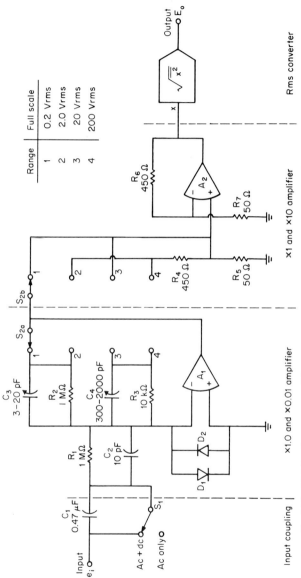

Fig. 4.15 Wideband ranging amplifier uses two stages to provide the optimum bandwidth and simple gain switching.

capacitors, and therefore gain switching in this stage is eased. The preamplifier uses A_1 in a closed-loop gain of 1 and 0.01, and the unity-gain, 100 percent feedback phase compensation is close to optimum for both conditions. The input impedance of the amplifier is R_1, and it is made 1 MΩ to minimize the source loading error; this high impedance level would prevent the preamplifier from having a very wide bandwidth, so the shunt capacitors C_2, C_3, and C_4 are included. The ratios of C_2 to C_3 and C_2 to C_4 must be adjustable to compensate for the input capacitance of the operational amplifier and the parasitic wiring capacitances. This compensation is most easily performed by making C_3 and C_4 adjustable and monitoring the output of the preamplifier with an oscilloscope; with a small-signal square-wave input, adjust the appropriate capacitor for a flat top on the output square wave.

The input to the preamplifier can be ac-coupled or dc-coupled by selecting the position of S_1. The value of the coupling capacitor is chosen so that the gain error at the lowest input frequency of interest is less than the overall accuracy required. If the maximum gain error allowed is K_ϵ, the value of the coupling capacitor for small errors should be

$$C_1 = \frac{1}{2\sqrt{2}\pi f_L R_1 \sqrt{K_\epsilon}}$$

If the input is ac-coupled, the offset voltages of the amplifiers appear as an uncorrelated input to the rms converter, and therefore add to the input signal in a square root of the sum of the squares fashion:

$$X_{rms} = A_{CL1} A_{CL2} \sqrt{E_{irms}^2 + V_{OS}^2}$$

where V_{OS} is the total equivalent input offset voltage. The result is that even relatively large input offset voltages produce little error in the output with the input ac-coupled. For the relatively large offset of 10 mV at the output of the X10 amplifier A_2, the error in the output of the rms converter is less than 0.5 mV with the minimum specified signal at this point of 100 mV rms. A 0.5 mV error at the output is an error of only 0.025 percent of full scale.

If the input is dc-coupled, the offset of the amplifiers must be kept small since it produces an error on the dc component of input signal in a linear fashion. However, wideband operational amplifiers typically have high voltage offset drift with time and temperature. Figure 4.16a and b shows two composite operational amplifiers.[5] Each utilizes a low-drift amplifier A_1 and a wideband amplifier A_2 in configurations which yield the best performance characteristics of each amplifier: the wide bandwidth of A_2 and the low offset voltage drift of A_1. An added benefit is increased open-loop gain. The low-drift operational amplifier provides continuous compensation for the input offset voltage of the fast amplifier. For the inverting

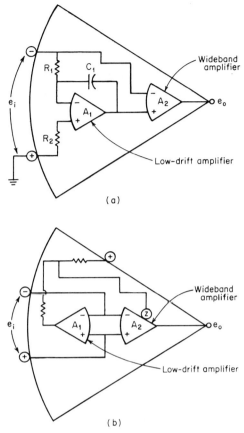

Fig. 4.16 Composite operational amplifiers can provide the best performance characteristics of both low-drift operational amplifiers and wide-bandwidth operational amplifiers.

composite amplifier shown in Fig. 4.16a, the low-drift operational amplifier A_1 senses the offset voltage from ground present at the summing junction of the wideband amplifier. Any such offset will be integrated by A_1 to develop an offset compensating voltage at the noninverting input of the fast amplifier. Integration continues until the summing junction voltage is offset from ground by only the input offset voltage of the low-drift amplifier (including the effects of the low-drift amplifier's input bias currents on resistors R_1 and R_2). Then the offset voltage of the composite amplifier is essentially that of the low-drift amplifier.

For the differential input composite operational amplifier shown in Fig. 4.16b, the low-drift amplifier amplifies the offset voltage at the inputs of the wideband amplifier and supplies an offset compensating voltage to the offset nulling terminal of the fast amplifier (pin Z of the fast amplifier). The output of the low-drift amplifier becomes whatever voltage is necessary to reduce the offset voltage of the fast amplifier to that of the low-drift amplifier. In other words, the offset voltage, and similarly the offset voltage drift, of the wide-bandwidth amplifier is again essentially replaced by that of the low-drift amplifier.

The open-loop gain of both composite amplifiers is $A_{02}(1 + \alpha A_{01})$, where α is defined as $e_x/A_1 e_i$. For the inverting-only composite operational amplifier, αA_1 is the response of the integrator formed by A_1, R_1, and C_1. For the differential input composite amplifier, α is essentially a constant, independent of frequency, and is the change in the offset voltage of A_2 due to a change in e_x divided by the change in e_x. It can be shown that for composite operational amplifiers to have a single pole in their open-loop gain response, $R_1 C_1$ must equal $A_{02}/2\pi f_{T2}$ in the inverting-only composite amplifier, and α must equal $-f_{T2}/A_{02} f_{T1}$ in the differential input composite amplifier, where f_{T1}, f_{T2}, and A_{02} are the unity-gain bandwidths of amplifiers A_1 and A_2 and the dc open-loop gain of A_2, respectively. In addition to a decreased offset voltage drift and an increased open-loop gain, the differential input composite amplifier has the further advantage of improved common-mode rejection at low frequencies. At low frequencies the common-mode rejection of the wide-bandwidth amplifier is essentially replaced by that of the low-drift amplifier, which is typically much better than that of the wideband amplifier.

Since the input impedance of the thermal rms converter is relatively low, 115 Ω for the thermoelement previously described, it may be necessary to include a power booster, or current booster, in series with the amplifier in the X10 stage of Fig. 4.15. Since such amplifiers are placed inside the feedback loop of the amplifier, they generally do not introduce any errors. They simply provide an increased output current capability. A further consideration in a complete rms-to-dc converter is the overall gain adjustment. It is possible to adjust the system gain by adjusting the value of the input ranging resistors in Fig. 4.15; however, this will interfere with the bandwidth compensation of C_3 and C_4. Also, adding additional wiring capacitance in these input stages is undesirable, and it is recommended that the system gain adjustment be performed at the output, dc end of the system, where it does not complicate the high-frequency performance. This may require separate gain adjustments for each range, which should be switched in tandem with the input ranging switch S_2 in Fig. 4.15. And a final note: Since the system is designed to allow measurement of up to 300 V rms inputs and the peak inputs may be much higher, input protection diodes D_1 and D_2 should be included to protect the input amplifier. These should be low-capacitance diodes.

4.4.3 High-crest-factor inputs

The low-frequency and high-frequency cutoffs for both the computing and thermal rms converters were discussed for sine-wave input signals in the previous sections. The results given for sine-wave inputs are generally adequate for most low-crest-factor inputs, such as practical sine-wave signals containing a few percent of distortion, dc plus sine-wave signals, typical noise signals encountered in practice, or even triangle waves. However, a square wave with a crest factor which is identical to that of a sine wave, 1.414, requires a 3 dB sine-wave bandwidth 100 times the square-wave frequency if 0.1 percent conversion accuracy is required. If the input signal is a pulse train having a 5:1 crest factor, only a 1 percent measurement accuracy will be obtained with a 3 dB sine-wave bandwidth 125 times the repetition rate of the pulse train. In addition to the high-frequency limitations which result from the large harmonic content in high-crest-factor signals, these signals also produce additional error at low frequencies. In Sec. 4.3.2 high-crest-factor errors with the thermal converter due to the voltage nonlinearity and temperature effects in the heater resistor were discussed. High-crest-factor errors with the computing rms converter also result because of nonideal log conformity in the logging-antilogging transistors.

Error analysis with nonsinusoidal inputs would be as varied as the possible input signals; however, a pulse train input is commonly used when discussing the high-crest-factor performance of an rms converter. While input signals which would require greater bandwidth and circuit linearity are theoretically possible, the pulse train represents a worst-case signal in a practical sense, and therefore the low-frequency and high-frequency requirements on the rms converter with high-crest-factor inputs will be discussed in terms of pulse train input signals. The rms value of a pulse train, such as the one shown in Fig. 4.17, can be determined from its mathematical definition, which is

$$E_{irms} = E_{ip} \sqrt{\frac{W}{P}}$$

Where E_{ip} is the peak value of the pulse train, W is the pulse width, and P is the signal period. If the pulse train is offset from zero by some dc level E_a, the rms value of the total signal is simply the square root of the sum of the squares of the two signals,

$$E_{irms} = \sqrt{E_a^2 + E_{ip}^2 \frac{W}{P}}$$

With sufficiently low repetition rates, the filtering inside the feedback loop of the rms converter will be inadequate, resulting in an error in the output, both an error in the output dc level and an instantaneous error due

170 Function Circuits

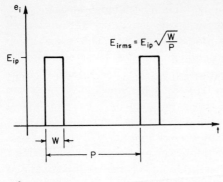

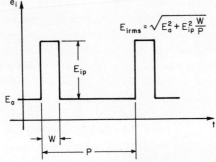

Fig. 4.17 Pulse train inputs are used to characterize the high-crest-factor performance of rms converters.

to the ripple. The low-frequency cutoff, or minimum repetition rate for a particular conversion accuracy, with pulse train input signals will be significantly higher than for sine-wave inputs. An approximate solution can be obtained by considering the case where the period of the input is small compared to the effective filter time constant τ_{eff}, given in Eq. (4–44) for the rms converters discussed. In Fig. 4.18 the input squared is shown both before and after low-pass filtering. The output is the square root of the dc level plus the ac signal. If the period of the pulse train is short compared to the effective filter time constant, the ac component can be approximated as a triangle wave. The peak-to-peak value of the triangle wave is $e_2^2 - e_1^2$, and for $P \ll \tau_{\text{eff}}$ the values of e_1^2 and e_2^2 can be determined from the approximate equations for the rise and fall,

$$e_2^2 \doteq e_1^2 + (E_{ip}^2 - e_1^2)\frac{W}{\tau_{\text{eff}}}$$

and

$$e_1^2 \doteq e_2^2 \left(1 - \frac{P - W}{\tau_{\text{eff}}}\right)$$

which yield the triangle-wave peak-to-peak value,

$$(e_2^2 - e_1^2) \doteq E_{ip}^2 \frac{W}{P} \frac{P - W}{\tau_{eff}}$$

Using a power series approximation for the square root yields the output,

$$e_o \doteq E_{dc} \left[1 + \frac{S(t)}{2E_{dc}^2} - \frac{S^2(t)}{8E_{dc}^4} \right]$$

where $S(t)$ is a triangle wave with a peak-to-peak amplitude of $(e_2^2 - e_1^2)$. The instantaneous error in the output voltage is the magnitude of the ripple, which is essentially due to the $S(t)$ term:

$$|e_{ripple}| \doteq \frac{E_{ip}}{4} \sqrt{\frac{W}{P}} \left(\frac{P - W}{\tau_{eff}} \right)$$

The $S^2(t)$ term has a dc component, $(e_2^2 - e_1^2)^2/12$, and therefore produces an error in the output dc level:

$$\epsilon_{odc} \doteq \frac{E_{ip}}{96} \sqrt{\frac{W}{P}} \left(\frac{P - W}{\tau_{eff}} \right)^2$$

Since $E_{irms} = E_{ip}(W/P)^{1/2}$, the output errors can be rewritten as

$$|e_{ripple}| \doteq \frac{E_{irms}}{4} \left(\frac{P - W}{\tau_{eff}} \right)$$

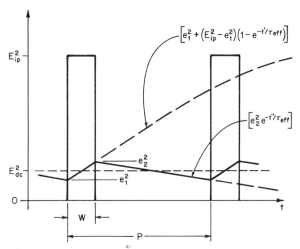

Fig. 4.18 After low-pass filtering, the input squared for a pulse train input has a dc component and an ac component. The ac component can be approximated as a triangle wave.

and
$$\epsilon_{odc} \doteq \frac{E_{irms}}{96}\left(\frac{P-W}{\tau_{eff}}\right)^2$$

For high crest factors, $P - W \approx P$ and the repetition rate is $1/P$; therefore, we can write the low-frequency cutoff for pulse train inputs f_{LP} as

$$f_{LP} \doteq \frac{1}{\tau_{eff}\sqrt{96K_{LP}}} \qquad P \gg W$$

where K_{LP} is the fractional error ϵ_{odc}/E_{irms} for pulse train inputs. Comparing this to Eqs. (4-11) and (4-36), we see that the low-frequency cutoff for pulse train inputs is about 5 times higher than the sine-wave low-frequency cutoff. With square-wave inputs where $P = W$ and the crest factor is 1.414, the low-frequency cutoff is still about 2.5 times higher than the sine-wave low-frequency cutoff.

With high-frequency pulse train input signals there are two primary limitations: the small-signal sine-wave bandwidth of the rms converter limits the number of harmonics which are measured, and the slew-rate limit of the input amplifier triangulates the input pulse, which reduces the energy content of the pulse. A pulse train input can be represented by a power series, and the Fourier series of the zero-based pulse train shown in Fig. 4.17 can be written as

$$e_i(t) = E_{ip}\frac{W}{P}\left[1 + 2\sum_{n=1}^{\infty}(-1)^n\frac{\sin nx}{nx}\cos\frac{2nxt}{W}\right] \qquad \text{where } x = \frac{\pi W}{P}$$

The rms value of the input is the rms sum of all the terms in the Fourier series,

$$E_{irms} = E_{ip}\frac{W}{P}\sqrt{1 + 2\sum_{n=1}^{\infty}\frac{\sin^2 nx}{(nx)^2}}$$

If the frequency response of the rms converter is limited, some of the harmonics will be excluded from the measurement. If the frequency response of the rms converter is approximated as being perfectly flat up to its 3 dB bandwidth and having zero response to signals above its 3 dB bandwidth, the output of the rms converter will be the truncated input,

$$E_o = E_{ip}\frac{W}{P}\sqrt{1 + 2\sum_{n=1}^{N}\frac{\sin^2 nx}{(nx)^2}}$$

where N is the number of harmonics measured. The number of harmonics measured is the ratio of the 3 dB bandwidth $f_{3\ dB}$ to the repetition frequency of the input f_p,

$$N = \frac{f_{3\ dB}}{f_P} = Pf_{3\ dB}$$

The truncation approximation of the rms converter's frequency response may seem crude; however, the energy content of the higher harmonics is small, and the unaccounted for loss of energy in the harmonics just below the −3 dB bandwidth is somewhat compensated by the partial accounting of the harmonics just above the −3 dB bandwidth.

The fractional error in the output K_{HP} due to limited bandwidth and a pulse train input is

$$K_{HP} = \frac{E_{irms} - E_o}{E_{irms}}$$

and then

$$K_{HP} = 1 - \sqrt{\frac{W}{P}\left[1 + 2\sum_{n=1}^{N} \frac{\sin^2 nx}{(nx)^2}\right]}$$

With the help of a computer, the output error with high-frequency pulse train inputs can be easily calculated using this equation. Figure 4.19 shows the results of this calculation; the percent error ($100 K_{HP}$ %) is plotted versus the pulse width–3 dB bandwidth product. The results of calculating the error for a symmetrical square wave (a square wave with zero

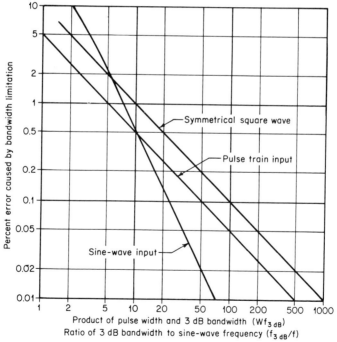

Fig. 4.19 Theoretical accuracy versus the frequency can be determined with this graph for pulse trains, square waves, and sine waves.

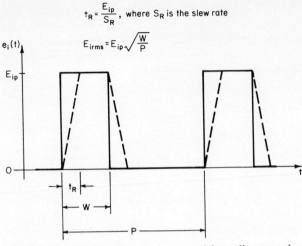

Fig. 4.20 Slew-rate limiting in the input amplifier will convert the input pulse train to a trapezoidal pulse train, which reduces the signals' energy content or rms value.

dc component) are also shown; exactly twice the error results for the same "pulse width." For comparison and reference, the theoretical error for sine-wave inputs is also shown. For sine waves the error is plotted versus the ratio of the converter's 3 dB bandwidth to the input signal frequency ($f_{3\text{ dB}}/f$).

It was stated earlier that the slew-rate limit of the input amplifier will triangulate the input pulse, which will reduce the energy content, and therefore the rms value, of the signal which is converted. The slew rate of the input amplifier required to obtain a particular conversion accuracy can be determined by computing the rms value of the trapezoidal pulse train which results. For simplicity it will be assumed that the input amplifier's slew rate is the same for positive or negative transitions. Further, it will be assumed that the slew rate is constant. While these assumptions may not be met with practical operational amplifiers, the result will be the worst-case error if the worst-case slew rate for the amplifier is used. Figure 4.20 shows the input pulse and the resulting slew-rate-limited pulse, whose rms value will be converted.

To determine the error in the output due to finite rise and fall times, we need to evaluate

$$\frac{1}{P} \int_0^P e^2(t)\, dt$$

This can be performed by breaking the integral into three parts,

$$E_o{}^2 = \frac{1}{P}\left[\int_0^{t_R} E_{ip}{}^2 \frac{t^2}{t_R{}^2}\,dt + \int_{t_R}^{W} E_{ip}{}^2\,dt + \int_{W}^{W+t_R} E_{ip}{}^2\left(1 - \frac{t-W}{t_R}\right)^2 dt\right]$$

$$= \frac{E_{ip}{}^2}{P}\left[\frac{t_R}{3} + (W - t_R) + \frac{t_R}{3}\right]$$

And since $E_{irms} = E_{ip}(W/P)^{1/2}$, the output can be written as

$$E_o = E_{irms}\sqrt{1 - \frac{t_R}{3W}}$$

The fractional error in the output due to finite slew rate K_{SR} is then

$$K_{SR} = 1 - \sqrt{1 - \frac{E_{ip}}{3WS_R}}$$

since

$$t_R = \frac{E_{ip}}{S_R}$$

where S_R is the amplifier's slew rate. For small errors (< 10%), $E_{ip}/3WS_R$ is much less than 1, allowing the expression for K_{SR} to be simplified to

$$K_{SR} \doteq \frac{E_{ip}}{6WS_R}$$

As an example, suppose the amplifier's slew rate is 100 V/μs and the peak input is 10 V; then the minimum pulse width for 1 percent accuracy is found to be 1.67 μs. If the pulse train has a crest factor of 5, the maximum repetition rate is 24 kHz. The small-signal 3 dB bandwidth of the amplifier, from Fig. 4.19, also must be greater than 3 MHz to ensure the 1 percent conversion accuracy.

From these results it is found that a typical computing rms converter with a sine-wave 3 dB bandwidth of 450 kHz will have a 0.1 percent bandwidth of 4.5 kHz when measuring simple square waves that have a crest factor of only 1.414. The upper repetition frequency for 1 percent measurement of a pulse train with a 5:1 crest factor will only be 3.6 kHz. The ultra-wide bandwidth of the thermal converter will provide a 0.1 percent accuracy measurement of a square wave to at least 1 MHz and 1 percent conversion of a 5:1 crest-factor pulse train with a repetition rate of 1 MHz. If the input heater resistor of the thermal converter is buffered by an amplifier, the resulting overall bandwidth may be reduced. However, even a general-purpose IC operational amplifier with a 1 MHz unity-gain bandwidth will provide a better overall frequency response than the computing rms converter.

REFERENCES

1. H. Handler and T. Cate, True RMS Voltage Conversion, *Electron. Des.*, Feb. 15, 1974.
2. J. G. Graeme, *Applications of Operational Amplifiers: Third-Generation Techniques*, McGraw-Hill Book Company, New York, 1973.
3. W. E. Ott, A New Technique of Thermal RMS Measurement, *IEEE J. Solid-State Circ.*, December 1974.
4. W. E. Ott, Monolithic Converter Augments AC-Measurement Capabilities, *Electronics*, Jan. 23, 1975.
5. W. E. Ott, Combined Op Amps Improve Overall Amplifier Response, *Electronics*, Nov. 8, 1973.

5

POWER SERIES AND FUNCTION GENERATORS

The term *function generator*, used in this chapter, should not be confused with the term *waveform generator*, commonly used in the electronics industry. The latter is a device generating a periodic waveform (square, triangular, pulses, etc.), while the former is an input-output block whose output voltage is a prescribed function of the input signal. The function may be given in equation form, in graphical form, or in a table of values. We will discuss a systematic method for approximating a given function and then modeling it with electronic analog-function circuits.

5.1 Power Series and Its Analog Realization[1,6]

Almost every function encountered in engineering applications can be represented by an infinite series. One of the most useful infinite series is the power series whose terms are arranged in ascending powers of x,

$$f(x) = a_0 + a_1 x + a_2 x^2 + a_3 x^3 + \cdots + a_n x^n + \cdots$$

The familiar Taylor's formula gives the power series expansion of an arbitrary function $f(x)$,

$$f(x) = f(a) + (x-a) f'(a) + \frac{(x-a)^2}{2!} f''(a) + \cdots + \frac{(x-a)^n}{n!} f^n(a) + \cdots$$

178 Function Circuits

In most applications, the first n terms of the power series are chosen to approximate $f(x)$ to the desired accuracy, and the $(n+1)$ and higher terms are discarded. Then the power series becomes a polynomial of the nth degree,

$$f(x) = a_0 + a_1 x + a_2 x^2 + \cdots + a_n x^n$$

This polynomial can be modeled by multipliers and operational amplifiers. The reader is referred to Chapter 1 regarding the fundamentals of operational amplifiers, especially Sec. 1.4.1 on summers. There are usually several different combinations of multipliers and operational amplifiers to implement the same polynomial. Figure 5.1 a to e shows the configura-

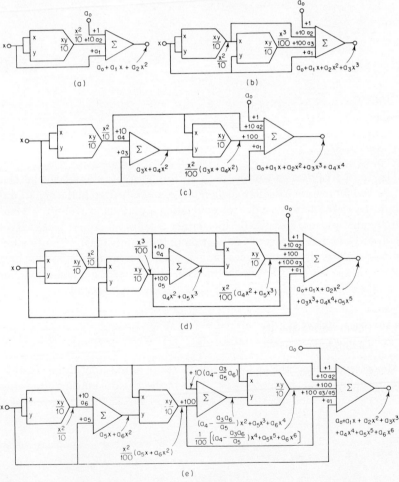

Fig. 5.1 (a) Realizing $f(x) = a_0 + a_1 + a_2 x^2$. (b) Realizing $f(x) = a_0 + a_1 x + a_2 x^2 + a_3 x^3$. (c) Realizing $f(x) = a_0 + a_1 x + a_2 x^2 + a_3 x^3 + a_4 x^4$. (d) Realizing $f(x) = a_0 + a_1 x + a_2 x^2 + a_3 x^3 + a_4 x^4 + a_5 x^5$. (e) Realizing $f(x) = a_0 + a_1 x + a_2 x^2 + a_3 x^3 + a_4 x^4 + a_6 x^6$.

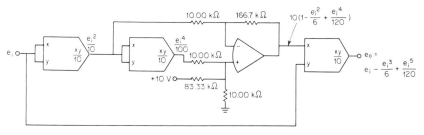

Fig. 5.2 Sine function generator approximating $\sin e_i = e_i + \dfrac{e_i^3}{3!} + \dfrac{e_i^5}{5!}$; less than 0.5 percent theoretical error in two quadrants between $-\pi/2$ and $+\pi/2$.

tions requiring the minimum number of multipliers to model a polynomial up to $n = 6$.

As an example, consider the sine function and its Taylor series,

$$\sin e_i = e_i - \frac{e_i^3}{3!} + \frac{e_i^5}{5!} - \frac{e_i^7}{7!} + \cdots$$

If the first three terms are retained, the fifth-degree polynomial approximates the sine function with less than 0.5 percent error for e_i between $-\pi/2$ and $+\pi/2$. Figure 5.2 shows a two-quadrant sine function generator using three multipliers and one operational amplifier.

By employing more terms of the Taylor series, better accuracies can be achieved at the expense of more components and more complex circuitry. The first four terms of Eq. (5-1) will provide 0.016 percent accuracy in the two quadrants between $-\pi/2$ and $+\pi/2$. In practice, this accuracy may not be achievable, since component tolerances would be the limiting factor.

5.2 Tchebysheff Polynomials[1,3]

In Sec. 5.1 we discussed the approximation of the sine function by expanding it into its Taylor's series. That method was straightforward to use. However, with the same number of components, the accuracy can be improved considerably if the same function is approximated by a series of Tchebysheff polynomials and only the first few terms are retained.

Tchebysheff polynomials are frequently used in engineering applications and are defined as

$$T_n(x) = \cos(n \cos^{-1} x) \quad \text{for } -1 < x < +1$$
$$T_n(x) = \cosh(n \cosh^{-1} x) \quad \text{for } |x| > 1$$

Hence for $n = 0$,

$$T_0(x) = 1$$

and for $n = 1$,

$$T_1(x) = x$$

TABLE 5.1 Coefficients of Tchebysheff Polynomials up to the Twentieth Degree.

T_n \ a_n	0	1	2	3	4	5	6	7	8	9	10	11	12	13	14	15	16	17	18	19	20
0	1	0	0	0	0	0	0	0	0	0	0	0	0	0	0	0	0	0	0	0	0
1	0	1	0	0	0	0	0	0	0	0	0	0	0	0	0	0	0	0	0	0	0
2	-1	0	2	0	0	0	0	0	0	0	0	0	0	0	0	0	0	0	0	0	0
3	0	-3	0	4	0	0	0	0	0	0	0	0	0	0	0	0	0	0	0	0	0
4	1	0	-8	0	8	0	0	0	0	0	0	0	0	0	0	0	0	0	0	0	0
5	0	5	0	-20	0	16	0	0	0	0	0	0	0	0	0	0	0	0	0	0	0
6	-1	0	18	0	-48	0	32	0	0	0	0	0	0	0	0	0	0	0	0	0	0
7	0	-7	0	56	0	-112	0	64	0	0	0	0	0	0	0	0	0	0	0	0	0
8	1	0	-32	0	160	0	-256	0	128	0	0	0	0	0	0	0	0	0	0	0	0
9	0	9	0	-120	0	432	0	-576	0	256	0	0	0	0	0	0	0	0	0	0	0
10	-1	0	50	0	-400	0	1,120	0	-1,280	0	512	0	0	0	0	0	0	0	0	0	0
11	0	-11	0	220	0	-1,232	0	2,816	0	-2,816	0	1,024	0	0	0	0	0	0	0	0	0
12	1	0	-72	0	840	0	-3,584	0	6,912	0	-6,144	0	2,048	0	0	0	0	0	0	0	0
13	0	13	0	-364	0	2,912	0	-9,984	0	16,640	0	-13,312	0	4,096	0	0	0	0	0	0	0
14	-1	0	98	0	-1,568	0	9,408	0	-26,880	0	39,424	0	-28,672	0	8,192	0	0	0	0	0	0
15	0	-15	0	560	0	-6,048	0	28,800	0	-70,400	0	92,160	0	-61,440	0	16,384	0	0	0	0	0
16	1	0	-128	0	2,688	0	-21,504	0	84,480	0	-180,224	0	212,992	0	-131,072	0	32,768	0	0	0	0
17	0	17	0	-816	0	11,424	0	-71,808	0	239,360	0	-452,608	0	487,424	0	-278,528	0	65,536	0	0	0
18	-1	0	162	0	-4,320	0	44,352	0	-228,096	0	658,944	0	-1,118,208	0	1,105,920	0	-589,824	0	131,072	0	0
19	0	-19	0	1,140	0	-20,064	0	160,512	0	-695,552	0	1,770,496	0	-2,723,840	0	2,490,368	0	-1,245,184	0	262,144	0
20	1	0	-200	0	6,600	0	-84,480	0	549,120	0	-2,050,048	0	4,659,200	0	-6,553,600	0	5,570,560	0	-2,621,440	0	524,288

It can be proved that
$$T_{n+1}(x) = 2T_n(x)T_1(x) - T_{n-1}(x)$$
Then
$$T_3(x) = 4x^3 - 3x$$
$$T_4(x) = 8x^4 - 8x^2 + 1$$
and so on.

Table 5.1 shows the coefficients of Tchebysheff polynomials in the general form
$$T_n(x) = a_0 + a_1 x + a_2 x^2 + \cdots + a_n x^n$$
up to $n = 20$.

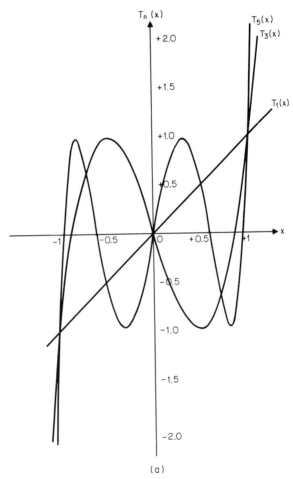

Fig. 5.3 Tchebysheff polynomials of (a) odd degrees $n = 1, 3,$ and 5, and (b) even degrees $n = 2$ and 4.

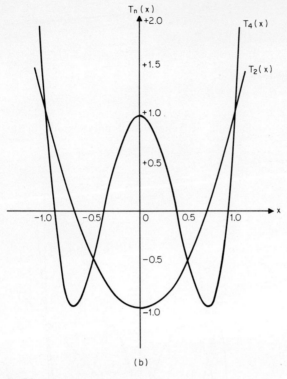

(b)

Fig. 5.3

Figure 5.3a shows a plot of $T_n(x)$ for $n = 1, 3, 5$, or odd-degree Tchebysheff polynomials, and Fig. 5.3b shows a plot for $n = 2, 4$, or even-degree Tchebysheff polynomials. The equal-ripple characteristic of $T_n(x)$, which will be used extensively in the following sections, is that it oscillates up and down with equal magnitude for x between -1 and $+1$.

5.2.1 Expansion into a series of Tchebysheff polynomials[1,3]

The orthogonality theory of Tchebysheff polynomials can be found in many advanced mathematics books and will not be discussed here. A practical example is used to illustrate the procedures for approximation with a series of Tchebysheff polynomials and to compare the improvement in accuracy over Taylor's series expansion.

Again consider the sine function and its Taylor's series,

$$\sin x = x - \frac{x^3}{3!} + \frac{x^5}{5!} - \frac{x^7}{7!} + \cdots$$

For engineering approximation purposes, only the first n terms will be considered, since the error from $(n + 1)$ and higher terms is negligible.

Then

$$\sin x = x - \frac{x^3}{3!} + \frac{x^5}{5!} - \frac{x^7}{7!} + \cdots \frac{x^n}{n!} \qquad (5\text{-}2)$$

The same function can be expanded into a series of Tchebysheff polynomials:

$$\sin x = a_0 T_0(x) + a_1 T_1(x) + a_2 T_2(x) + \cdots + a_n T_n(x) \qquad (5\text{-}3)$$

To achieve high accuracy, n should be chosen as large as is practical, depending on whether a computer or a desk-top calculator is available to do the computation. For illustration purposes, make $n = 5$. From Eq. (5-2),

$$\sin x = x - \frac{x^3}{3!} + \frac{x^5}{5!} \qquad (5\text{-}4)$$

For a two-quadrant sine-function generator, x ranges from $-\pi/2$ to $+\pi/2$. It has to be rescaled to range from -1 to $+1$ in order to utilize the equal-ripple characteristics of Tchebysheff polynomials. Assign

$$y = \frac{2}{\pi} x \quad \text{or} \quad x = \frac{\pi}{2} y \qquad (5\text{-}5)$$

Substitute Eq. (5-5) into (5-4).

$$\sin \frac{\pi}{2} y = \frac{\pi}{2} y - \frac{1}{3!}\left(\frac{\pi}{2}\right)^3 y^3 + \frac{1}{5!}\left(\frac{\pi}{2}\right)^5 y^5 \qquad (5\text{-}6)$$

Substitute Eq. (5-5) into (5-3) with $n = 5$.

$$\sin \frac{\pi}{2} y = a_0 T_0(y) + a_1 T_1(y) + a_2 T_2(y) + a_3 T_3(y) + a_4 T_4(y) + a_5 T_5(y) \qquad (5\text{-}7)$$

From Table 5.1,

$$T_0(y) = 1 \qquad (5\text{-}8)$$
$$T_1(y) = y \qquad (5\text{-}9)$$
$$T_2(y) = 2y^2 - 1 \qquad (5\text{-}10)$$
$$T_3(y) = 4y^3 - 3y \qquad (5\text{-}11)$$
$$T_4(y) = 8y^4 - 8y^2 + 1 \qquad (5\text{-}12)$$
$$T_5(y) = 16y^5 - 20y^3 + 5y \qquad (5\text{-}13)$$

Substitute Eqs. (5-8) through (5-13) into (5-7) and then equate coefficients of the same y-degree terms with Eq. (5-6).

$$y^5: \quad 16a_5 = \frac{1}{5!}\left(\frac{\pi}{2}\right)^5 \qquad (5\text{-}14)$$

$$y^4: \quad 8a_4 = 0 \qquad (5\text{-}15)$$

$$y^3: \quad -20a_5 + 4a_3 = -\frac{1}{3!}\left(\frac{\pi}{2}\right)^3 \qquad (5\text{-}16)$$

$$y^2: \quad -8a_4 + 2a_2 = 0 \qquad (5\text{-}17)$$

$$y^1: \quad 5a_5 - 3a_3 + a_1 = \frac{\pi}{2} \qquad (5\text{-}18)$$

$$y^0: \quad a_4 - a_2 + a_0 = 0 \qquad (5\text{-}19)$$

Solve Eqs. (5-14) through (5-19), starting always with the highest degree first.

$$a_5 = 0.00498077$$
$$a_4 = 0$$
$$a_3 = -0.136587$$
$$a_2 = 0$$
$$a_1 = 1.13614$$
$$a_0 = 0$$

Rewrite Eq. (5-7) with known coefficients.

$$\sin \frac{\pi}{2} y = 1.13614 T_1(y) - 0.136587 T_3(y) + 0.00498077 T_5(y)$$

TABLE 5.2 Approximation of sin x by Truncated Taylor's Series and by Truncated Tchebysheff Polynomials.

x (degree)	x (radians)	$\sin x$	$x - 0.166667 x^3$	$0.984151 x - 0.140965 x^3$
-90	-1.57080	-1.00000	-0.924831	-0.999551
-80	-1.39626	-0.984808	-0.942582	-0.990414
-70	-1.22173	-0.939693	-0.917799	-0.945305
-60	-1.04720	-0.886025	-0.855802	-0.868720
-50	-0.872664	-0.766044	-0.761902	-0.765152
-40	-0.698131	-0.642788	-0.641421	-0.639102
-30	-0.523598	-0.500000	-0.499674	-0.495064
-20	-0.349066	-0.342020	-0.341977	-0.337538
-10	-0.174533	-0.173648	-0.173647	-0.171017
0	0	0	0	0
$+10$	$+0.174533$	$+0.173648$	$+0.173647$	$+0.171017$
$+20$	$+0.349066$	$+0.342020$	$+0.341977$	$+0.337538$
$+30$	$+0.523598$	$+0.500000$	$+0.499674$	$+0.495064$
$+40$	$+0.698131$	$+0.642788$	$+0.641421$	$+0.639102$
-50	$+0.872664$	$+0.766044$	$+0.761902$	$+0.765152$
$+60$	$+1.04720$	$+0.886025$	$+0.855802$	$+0.868720$
$+70$	$+1.22173$	$+0.939693$	$+0.917799$	$+0.945305$
$+80$	$+1.39626$	$+0.984808$	$+0.942582$	$+0.990414$
$+90$	$+1.57080$	$+1.00000$	$+0.924831$	$+0.999551$

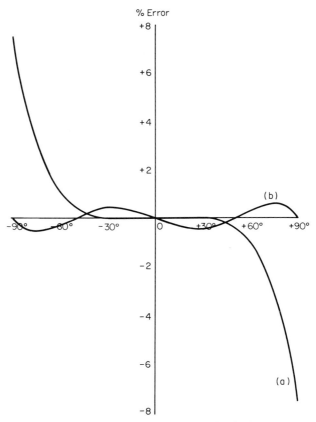

Fig. 5.4 Error curves of sin x approximation (a) by the first two terms of Taylor's series expansion, $x - 0.166667x^3$; (b) by truncated Tchebysheff polynomials, $0.984151x - 0.140965x^3$.

Now retain only the first two terms.

$$\sin \frac{\pi}{2} y = 1.13614 T_1(y) - 0.136587 T_3(y) \quad (5\text{--}20)$$

Substitute Eqs. (5-9) and (5-11) into (5-20) and simplify.

$$\sin \frac{\pi}{2} y = 1.54590 y - 0.546348 y^3 \quad (5\text{--}21)$$

Change the y variable into x by substituting Eq. (5-5) into (5-21).

$$\sin x \doteq 0.984151x - 0.140965x^3 \quad (5\text{--}22)$$

For comparison, retain the first two terms of the Taylor's expansion in Eq. (5-2).

$$\sin x \doteq x - 0.166667x^3 \qquad (5\text{-}23)$$

Equation (5-22) is the approximation of $\sin x$ by Tchebysheff polynomials truncated to two terms, and (5-23) is the approximation by Taylor's expansion, also truncated to two terms. Their respective coefficients are quite different. In Table 5.2, the theoretical values of $\sin x$ are listed in the third column and the approximations by Taylor's and Tchebysheff's expansions are listed in the fourth and fifth columns, respectively. A plot of the error curves for these two approximations is shown in Fig. 5.4. Notice that Taylor's expansion provides excellent accuracy for x around zero, but the error increases to 7.5 percent as $|x|$ approaches 90°. However, the Tchebysheff expansion provides equal-ripple error of less than 0.6 percent across the whole range of x from $-90°$ to $+90°$.

5.2.2 Computer-aided expansion into Tchebysheff polynomials[3,7]

In the $\sin x$ approximation discussed in the previous section, $n = 5$ was chosen to simplify computation. To approximate more complex functions and still preserve acceptable accuracy, n should be made as large as possible, which means a computer program is needed to do the computations. A recursive formula will be derived, and a program written in Fortran language will use it to expand any function, given by its power series, into a new series of Tchebysheff polynomials.

Consider any arbitrary function $f(x)$ given by

$$f(x) = a_0 + a_1 x + a_2 x^2 + \cdots + a_n x^n$$

Since the Fortran language used by most computers will not recognize a_0, $f(x)$ has to be rewritten starting with a_1.

$$f(x) = a_1 + a_2 x + a_3 x^2 + \cdots + a_n x^{n-1} \qquad (5\text{-}24)$$

The last term $a_{n+1}x^n$ is dropped to simplify the computer program without sacrificing the accuracy, since n is an input constant to the program.

Expand $f(x)$ into a series of Tchebysheff polynomials.

$$f(x) = b_1 + b_2 T_1(x) + b_3 T_2(x) + \cdots + b_n T_{n-1}(x) \qquad (5\text{-}25)$$

where $T_1(x) = C_{1,1} + C_{1,2}x$
$T_2(x) = C_{2,1} + C_{2,2}x + C_{2,3}x^2$
$T_3(x) = C_{3,1} + C_{3,2}x + C_{3,3}x^2 + C_{3,4}x^3$
.
.
.
$T_{n-1}(x) = C_{n-1,1} + C_{n-1,2}x + C_{n-1,3}x^2 + \cdots + C_{n-1,n}x^{n-1}$

Equate coefficients of the same x degree in Eqs. (5–24) and (5–25), starting with x^{n-1} first. Then

$$a_n = C_{n-1,n}b_n$$
$$a_{n-1} = C_{n-1,n-1}b_n + C_{n-2,n-1}b_{n-1}$$
$$a_{n-2} = C_{n-1,n-2}b_n + C_{n-2,n-2}b_{n-1} + C_{n-3,n-2}b_{n-2}$$

.
.
.

$$a_{n-i} = C_{n-1,n-i}b_n + C_{n-2,n-i}b_{n-1} + C_{n-3,n-i}b_{n-2} + \cdots$$
$$+ C_{n-i,n-i}b_{n-i+1} + C_{n-i+1,n-i}b_{n-i} \quad (5\text{–}26)$$

Rewrite Eq. (5–26) in closed form.

$$a_{n-i} = \sum_{j=1}^{i} C_{n-j,n-i}b_{n-j+1} + C_{n-i-1,n-i}b_{n-i} \quad (5\text{–}27)$$

From Eq. (5–27), the recursive formula for b_{n-i} is

$$b_{n-i} = \frac{a_{n-i} - \sum_{j=1}^{i} C_{n-j,n-i}b_{n-j+1}}{C_{n-i-1,n-i}} \quad (5\text{–}28)$$

where $i = 1, n$

Program TCHEB is listed in Fig. 5.5. The program consists mainly of four parts.

1. Constants of Tchebysheff polynomials up to T_{20}, from Table 5.1, are stored as data statements.
2. The program will ask for input data, n, $a_1, a_2, a_3, \ldots, a_n$, of the given power series up to the x^{n-1} term.
3. It then computes $b_1, b_2, b_3, \ldots, b_n$ of the series of Tchebysheff polynomials, using the recursive formula in Eq. (5–28).
4. At last the program prints out both the input and output data.

As an example, the same sine function is used to run the program. The output data are presented in Fig. 5.6. Twelve terms are used for approximation, so $n = 12$. The column under A(I) lists the first 12 coefficients of $\sin x$ in its Taylor's series. Note that x is scaled from $\pm \pi/2$ to ± 1; hence a_2 is $\pi/2$ instead of 1 and a_4 is $-(\pi/2)^3/3!$ instead of the regular $-1/3!$ and so on. Twelve coefficients of the output series of Tchebysheff polynomials are listed in the third column under B(I).

If we retain the first four terms or the first two nonzero terms and compare b_2 and b_4 with the corresponding coefficients in Eq. (5–20), the improvement in accuracy is noticeable, but there is not much improvement

```
00100       PROGRAM TCHEB(INPUT,OUTPUT)
00105C      GENERATE COEFFICIENTS OF TCHEBYSCHEFF POLYNOMIALS
00106C      B(1),(B2), . . . , B(N) IN
00107C      F(X)=B(1)+B(2)*T1(X)+B(3)*T2(X)+ . . .+B(N)*TN1(X),N1=N-1
00108C      GIVEN F(X)=A(1)+A(2)*X+A(3)*X**2+ . . .+A(N)*X**(N-1)
00110       DIMENSION A(20),B(20),C1(20),C2(20),C3(20),C4(20)
00120       DIMENSION C5(20),C6(20),C7(20),C8(20),C9(20),C10(20)
00130       DIMENSION C11(20),C12(20),C13(20),C14(20),C15(20)
00140       DIMENSION C16(20),C17(20),C18(20),C19(20),C20(20)
00150       DIMENSION C(20,21),CC(20)
00155C      STORE DATA FOR CONSTANTS OF TCHEBYSCHEFF POLYNOMIALS
00156C      UP TO 20TH ORDER
00160       DATA((C1(I),I=1,20)=0.,1.,18(0.))
00170       DATA((C2(I),I=1,20)=-1.,0.,2.,17(0.))
00180       DATA((C3(I),I=1,20)=0.,-3.,0.,4.,16(0.))
00190       DATA((C4(I),I=1,20)=1.,0.,-8.,0.,8.,15(0.))
00200       DATA((C5(I),I=1,20)=0.,5.,0.,-20.,0.,16.,14(0.))
00210       DATA((C6(I),I=1,20)=-1.,0.,18.,0.,-48.,0.,32.,13(0.))
00220       DATA((C7(I),I=1,20)=0.,-7.,0.,56.,0.,-112.,0.,
00230+      64.,12(0))
00240       DATA((C8(I),I=1,20)=1.,0.,-32.,0.,160.,0.,-256.,
00250+      0.,128.,11(0.))
00260       DATA((C9(I),I=1,20)=0.,9.,0.,-120.,0.,432.,0.,-576.,
00270+      0.,256.,10(0.))
00280       DATA((C10(I),I=1,20)=-1.,0.,50.,0.,-400.,0.,1120.,
00290+      0.,-1280.,0.,512.,9(0.))
00300       DATA(( CC11(I),I=1,20)=0.,-11..0.,220.,0.,-1232.,0.,
00310+      2816.,0.,-2816.,0.,1024.,8(0.))
00320       DATA((C12(I),I=1,20)=1.,0.,-72.,0.,840.,0.,-3584.,
00330+      0.,6912.,0.,-6144.,0.,2048.,7(0.))
00340       DATA((C13(I),I=1,20)=0.,13.,0.,-364.,0.,2912.,0.,
00350+      -9984.,0.,16640.,0.,-13312.,0.,4096.,6(0.))
00360       DATA((C14(I),I=1,20)=-1.,0.,98.,0.,-1568.,0.,9408.,
00370+      0.,-26880.,0.,39424.,0.,-28672.,0.,8192.,5(0.))
00380       DATA((C15(I),I=1,20)=0.,-15.,0.,560.,0.,-6048.,0.,
00390+      28800.,0.,-70400.,0.,92160.,0.,-61440.,0.,16384.,4(0.))
00400       DATA((C16(I),I=1,20)=1.,0.,-128.,0.,2688.,0.,-21504.,
00410+      0.,84480.,0.,-180224.,0.,212992.,0.,-131072.,
00420+      0.,32768.,3(0.))
00430       DATA((C17(I),I=1,20)=0.,17.,0.,-816.,0.,11424.,0.,
00440+      -71808.,0.,239360.,0.,-452608.,0.,487424.,0.,
00450+      -278528.,0.,65536.,2(0.))
00460       DATA((C18(I),I=1,20)=-1.,0.,162.,0.,-4320.,0.,44352.,
00470+      -2288096.,0.,658944.,0.,-1118208.,0.,1105920.,0.,
00480+      -589824.,0.,131072.,0.)
00490       DATA((C19(I),I=1,20)=0.,-19.,0.,1140.,0.,-20064.,0.,
00500+      160512.,0.,-695552.,0.,1770496.,0.,-2723840.,0.,
00510+      2490368.,0.,-1245184.,0.,262144)
00512       DATA((C20(I),I=1,20)=1.,0.,-200.,0.,6600.,0.,-84480.,
00513+      0.,549120.,0.,-2050048.,0.,4659200.,0.,-6553600.,
00514+      0.,5570560.,0.,-2621440.,0.)
00515       C(20,21)=524288.
00540       DO 5 I=1,20
```

Fig. 5.5 Generate a series of Tchebysheff polynomials $f(x) = b_1 + b_2 T_1(x) + b_3 T_2(x) + \cdots + b_n T_{n-1}(x)$, given $f(x) = a_1 + a_2 x + a_3 x^2 + \cdots + a_n x^{n-1}$.

```
00550      C(1,I)=C1(I)
00560      C(2,I)=C2(I)
00570      C(3,I)=C3(I)
00580      C(4,I)=C4(I)
00590      C(5,I)=C5(I)
00595      C(6,I)=C6(I)
00600      C(7,I)=C7(I)
00610      C(8,I)=C8(I)
00620      C(9,I)=C9(I)
00630      C(10,I)=C10(I)
00640      C(11,I)=C11(I)
00650      C(12,I)=C12(I)
00660      C(13,I)=C13(I)
00670      C(14,I)=C14(I)
00680      C(15,I)=C15(I)
00690      C(16,I)=C16(I)
00700      C(17,I)=C17(I)
00710      C(18,I)=C18(I)
00720      C(19,I)=C19(I)
00730      C(20,I)=C20(I)
00740      5 CONTINUE
00750C     ENTER INPUT DATA, NO OF TERMS N, A(1), A(2), . . . , A(N)
00770      READ 20,N
00780      20 FORMAT(I2)
00790      DO 40 I = 1,N
00820      READ 30, A(I)
00830      30 FORMAT(E14.6)
00840      40 CONTINUE
00845C     COMPUTE B(1),B(2), . . . , B(N)
00850      N2=N−1
00860      B(N)=A(N)/C(N−1,N)
00870      DO 70 I=1,N2
00880      DO 50 J=1,I
00890      CC(J)=C(N−J,N−I)*B(N−J+1)
00900      50 CONTINUE
00910      DO 60 J=1,I
00915      IF(J−1) 60,60,65
00920      65 CC(J)=CC(J)+CC(J−1)
00930      60 CONTINUE
00935      IF(I−N+1) 100,110,110
00940      100 B(N−I)=(A(N−I)−CC(I))/C(N−I−1,N−I)
00950      70 CONTINUE
00960      110 B(1)=A(1)−CC(N−1)
00962C     PRINT OUT INPUT DATA A(1), A(2), . . . , A(N) AND
00964C     OUTPUT DATA B(1),B(2), . . . , B(N)
00967      PRINT 120
00968      120 FORMAT(6X,*I*,10X,*A(I)*,16X,*B(I)*)
00971      DO 80 I=1,N
00980      PRINT 90,I,A(I),B(I)
00990      90 FORMAT(5X,I2,5X,E14.6,5X,E14.6)
01000      80 CONTINUE
01010      RETURN
01020      END
```

190 Function Circuits

I	A(I)	B(I)
1	0.	0.
2	1.570800E+00	1.133656E+00
3	0.	0.
4	-6.459620E-01	-1.380690E-01
5	0.	0.
6	7.969230E-02	4.491846E-03
7	0.	0.
8	-4.681730E-03	-6.731827E-05
9	0.	0.
10	1.604400E-04	6.653778E-07
11	0.	0.
12	3.598810E-06	3.514463E-09

Fig. 5.6 Approximation of sin x using program TCHEB. Column A(I) lists input data, Taylor's coefficients. Column B(I) lists output data, Tchebysheff coefficients.

because the Taylor's series of sin x converges relatively fast. Rescaling b_2 and b_4 back to the angular domain of $\pm \pi/2$ gives

$$\sin x \doteq 0.985399x - 0.142496x^3 \qquad (5\text{-}29)$$

This approximation of sin x will yield less than 0.45 percent error from $-\pi/2$ to $+\pi/2$. A two-quadrant sine generator using two multipliers and one operational amplifier is shown in Fig. 5.7.

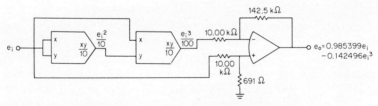

Fig. 5.7 Two-quadrant sine function generator approximating $\sin e_i = 0.985399e_i - 0.142496e_i^3$, with a theoretical error of less than 0.45 percent.

5.3 Arbitrary Function Generators

The last section presented a systematic method for designing an analog-function generator whose function was given by a power series. Quite often in engineering applications, a function is given in graphical form or as a table of numbers.

Assume a curve is given in graphical form as shown in Fig. 5.8. Enough points along the curve can be picked to accurately describe $y(e)$, i.e., $e_1, y_1; e_2, y_2; e_3, y_3; \ldots; e_n, y_n$. Naturally, more points should be picked along the portion where the slope of the curve changes fast. In order to make use of Tchebysheff polynomials and to minimize the error function, the two end points e_1 and e_n should be scaled to -1 and $+1$, respectively.

Let's use a new variable x such that $x_1 = -1$ and $x_n = +1$; hence

$$x = \frac{2e - e_n - e_1}{e_n - e_1} \tag{5-30}$$

The original curve $y(e)$ will temporarily become $y(x)$ and the set of points become $-1, y_1; x_2, y_2; x_3, y_3; \ldots; x_{n-1}, y_{n-1}; +1, y_n$.

Assume $y(x)$ can be formulated by a polynomial of the nth degree,

$$y(x) = a_0 + a_1 x + a_2 x^2 + \cdots + a_n x^n \tag{5-31}$$

Substitute the values of each point x_i, y_i for $i = 1, n$ into Eq. (5-31).

$$y_1 = a_0 + a_1(-1) + a_3(-1)^2 + \cdots + a_n(-1)^n$$
$$y_2 = a_0 + a_1 x_2 + a_2 x_2^2 + \cdots + a_n x_2^n$$
$$y_3 = a_0 + a_1 x_3 + a_2 x_3^2 + \cdots + a_n x_3^n$$
$$\vdots$$
$$y_{n-1} = a_0 + a_1 x_{n-1} + a_2 x_{n-1}^2 + \cdots + a_n x_{n-1}^n$$
$$y_n = a_0 + a_1(+1) + a_2(+1)^2 + \cdots + a_n(+1)^n$$

Now we have n simultaneous linear equations and n unknowns, $a_0, a_1, \ldots, a_n$. They can be solved for by the gaussian elimination method. Substituting the solved values of $a_0, a_1, \ldots, a_n$ back into Eq. (5-31), an nth-degree polynomial is established to describe the curve originally given by either a graph or a table of numbers.

n usually is very high, since in most cases 10 to 20 points are needed to represent the curve. It is too expensive and impractical to electronically simulate such a high-degree polynomial. The numerical method in Sec. 5.2.1 or the computer program in Sec. 5.2.2 should always be employed to generate a new series of Tchebysheff polynomials and truncate it to the first few terms.

The Taylor's series discussed in the previous sections are always con-

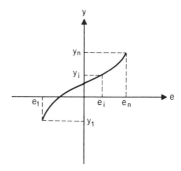

Fig. 5.8 Enough points along the curve, especially where the slope changes fast, should be picked to accurately describe the function $y(e)$.

vergent, that is, $a_0 > a_1 > a_2 > \cdots > a_n > a_{n+1} \cdots$, so the error in the polynomial retained from the first $(n-1)$ terms will differ insignificantly from that from n terms so long as n is large. However, the polynomial in Eq. (5-31) may not be convergent, and hence none of the terms can be omitted.

5.3.1 Computer-aided n-degree polynomial generation[7]

The Fortran program POWSERS listed in Fig. 5.9 will do the complex computation required to solve n linear simultaneous equations, up to $n = 20$. The program will ask for input data n (number of points), $x_1, x_2, \ldots, x_n$ and $y_1, y_2, \ldots, y_n$. Coefficients $a_0, a_1, \ldots, a_n$ in Eq. (5.31) will be printed out as output data. Again, since the Fortran language will not recognize a_0, the actual printed-out polynomial is

$$y(x) = a_1 + a_2 x + a_3 x^2 + \cdots + a_n x^{n-1}$$

As an example, the same two-quadrant sine function is illustrated. This time it is assumed that the Taylor's series for the sine function is not known, but that the function's characteristic is defined by 19 equally spaced points from $-90°$ to $+90°$, as listed in the first two columns of Table 5.3. Θ is converted to radians and listed in the third column. It is then scaled to the normalized range from -1 to $+1$ through variable transformations from θ to x such that

$$x = \frac{\theta}{1.57080} \tag{5-32}$$

The values of x and y are entered into the program POWSERS, and the output data are printed in Fig. 5.10. The values listed in the column under A(I) are the coefficients of the eighteenth-degree polynomial,

$$y(x) = a_1 + a_2 x + a_3 x^2 + \cdots + a_{19} x^{18}$$

Notice that $a_1, a_3, a_5, a_7, \ldots, a_{19}$ are so small that they can be considered zeros. Hence,

$$y(x) = a_2 x + a_4 x^3 + a_6 x^5 + \cdots + a_{18} x^{17}$$

where $a_2 = 1.570793$, $a_4 = -0.6456169$, and so on, as listed in Fig. 5.10.

Now recall program TCHEB in Fig. 5.5. Enter $a_1, a_2, \ldots, a_{18}$ as its input data. Notice that only 18 coefficients are entered; $a_{19} = 0$ is deleted because the computer will overflow if the last coefficient is read zero. The output data are printed in Fig. 5.11, where B(I) are the coefficients of a series of Tchebysheff polynomials,

$$y(x) = b_1 + b_2 T_1(x) + b_3 T_2(x) + \cdots + b_{18} T_{17}(x)$$

```
100      PROGRAM POWSERS(INPUT,OUTPUT)
110C     GENERATE Y(X)=A(1)+A(2)*X+A(3)*X**2+...+A(N)*X**(N-1)
120C     GIVEN X(1),Y(1);X(2),Y(2);...; X(N),Y(N)
130      DIMENSION X(20),Y(20),Z(20,20)
140C     ENTER INPUT DATA, NO. OF POINTS N, X(1),X(2), ..., X(N)
150C     AND Y(1),Y(2), ..., Y(N)
160      READ 10,N
170      10 FORMAT(I2)
180      DO 20 I=1,N
190      20 READ 30,X(I)
200      30 FORMAT(E14.6)
210      DO 40 I=1,N
220      40 READ 50,Y(I)
230      50 FORMAT(E14.6)
240C     PRINT OUT INPUT DATA X(1),Y(1);X(2),Y(2);...; X(N),Y(N)
250      PRINT 60
260      60 FORMAT(//,6X,*N*,10X,*X(I)*,16X,*Y(I)*)
270      DO 70 I=1,N
280      70 PRINT 80,I,X(I),Y(I)
290      80 FORMAT(5X,I2,5X,E14.6,5X,E14.6)
300C     GENERATE COEFFICIENTS FOR N SIMULTANEOUS LINEAR
310C     EQUATIONS
320      DO 90 I=1,N
330      90 Z(I,1)=1.
340      DO 100 I1=1,N
350      DO 110 I2=2,N
360      Z(I1,I2)=X(I1)**(I2-1)
370      110 CONTINUE
380      100 CONTINUE
390C     SOLVE N SIMULTANEOUS LINEAR EQUATIONS
400      DO 120 I1=1,N
410      W=0.
420      DO 140 I2=I1,N
430      IF(ABS(Z(I2,I1))-W) 140,150,150
440      150 W=ABS(Z(I2,I1))
450      I1=I2
460      140 CONTINUE
470      IF(I1-I2)160,170,180
480      180 DO 190 I2=1,N
490      W=Z(I1,I2)
500      Z(I1,I2)=Z(II,I2)
510      190 Z(II,I2)=W
520      W=Y(I1)
530      Y(I1)=Y(II)
540      Y(II)=W
550      170 W=Z(I1,I2)
560      DO 200 I2=I1,N
570      200 Z(I1,I2)=Z(I1,I2)/W
580      Y(I1)=Y(I1)/W
590      IF(I1-N)210,220,160
600      210 N1=I1+1
```

Fig. 5.9 Program to generate $y = a_1 + a_2 x + a_3 x^2 + \cdots + a_n x^{n-1}$, given n points of an arbitrary curve, $x_1 y_1; x_2 y_2; \ldots ; x_n y_n$.

```
610      DO 230 I2=N1,N
620      W=Z(I2,I1)
630      DO 240 I3=N1,N
640  240 Z(I2,I3)=Z(I2,I3)-W*Z(I1,I3)
650      Y(I2)=Y(I2)-W*Y(I1)
660  230 CONTINUE
670  120 CONTINUE
680  220 IF(N-1)160,160,250
690  250 N1=N
700  260 DO 270 I2=N1,N
710  270 Y(N1-1)=Y(N1-1)-Y(I2)*Z(N1-1,I2)
720      N1=N1-1
730      IF(N1-1)160,160,260
740C     PRINT OUT OUTPUT DATA A(1),A(2),..., A(N)
750  160 PRINT 280
760  280 FORMAT(//,6X,*N*,12X,*A(I)*)
770      DO 290 I=1,N
780  290 PRINT 300,I,Y(I)
790  300 FORMAT(5X,I2,5X,E15.6)
800      RETURN
810      END
```

Fig. 5.9

TABLE 5.3 Sine Function Given by a Table of Values. θ Is Scaled to -1 and $+1$ by $x = \theta/1.57080$.

θ	y	θ(radians)	$x = \theta/1.57080$
$-90°$	-1.00000	-1.57080	-1.00000
$-80°$	-0.984808	-1.39626	-0.888887
$-70°$	-0.939693	-1.22173	-0.777778
$-60°$	-0.866025	-1.04720	-0.666668
$-50°$	-0.766044	-0.872665	-0.555556
$-40°$	-0.642788	-0.698132	-0.444445
$-30°$	-0.500000	-0.523599	-0.333334
$-20°$	-0.342020	-0.349066	-0.222222
$-10°$	-0.173648	-0.174533	-0.111111
$0°$	0	0	0
$+10°$	$+0.173648$	$+0.174533$	$+0.111111$
$+20°$	$+0.342020$	$+0.349066$	$+0.222222$
$+30°$	$+0.500000$	$+0.523599$	$+0.333334$
$+40°$	$+0.642788$	$+0.698132$	$+0.444445$
$+50°$	$+0.766044$	$+0.872665$	$+0.555556$
$+60°$	$+0.866025$	$+1.04720$	$+0.666668$
$+70°$	$+0.939693$	$+1.22173$	$+0.777778$
$+80°$	$+0.984808$	$+1.39626$	$+0.888887$
$+90°$	$+1.00000$	$+1.57080$	$+1.00000$

N	X(I)	Y(I)
1	-1.000000E+00	-1.000000E+00
2	-8.888870E-01	-9.848080E-01
3	-7.777780E-01	-9.396930E-01
4	-6.666680E-01	-8.660250E-01
5	-5.555560E-01	-7.660440E-01
6	-4.444450E-01	-6.427880E-01
7	-3.333340E-01	-5.000000E-01
8	-2.222220E-01	-3.420200E-01
9	-1.111110E-01	-1.736480E-01
10	0.	0.
11	1.111110E-01	1.736480E-01
12	2.222220E-01	3.420200E-01
13	3.333340E-01	5.000000E-01
14	4.444450E-01	6.427880E-01
15	5.555560E-01	7.660440E-01
16	6.666680E-01	8.660250E-01
17	7.777780E-01	9.396930E-01
18	8.888870E-01	9.848080E-01
19	1.000000E+00	1.000000E+00

N	A(I)
1	2.756717E-14
2	1.570793F+00
3	-3.229334E-13
4	-6.456169E-01
5	1.187780E-11
6	7.102730E-02
7	-1.547836E-10
8	7.896531E-02
9	1.029481E-09
10	-3.987088E-01
11	-3.930006E-09
12	1.028900E+00
13	8.874419E-09
14	-1.456586E+00
15	-1.159924E-08
16	1.060440E+00
17	8.045581E-09
18	-3.092132E-01
19	-2.277006E-09

Fig. 5.10 Generate an eighteenth-degree polynomial to approximate the sine function given by 19 sets of x, y values.

If the series is truncated to the first two nonzero terms, then

$$y(x) = 1.133697 T_1(x) - 0.1380333 T_3(x) \qquad (5\text{-}33)$$

Substitute $T_1(x)$ and $T_3(x)$ from Table 5.1 into Eq. (5-33).

$$y(x) = 1.54780x - 0.552133x^3 \qquad (5\text{-}34)$$

Change x back to the original variable θ in radians by substituting Eq. (5-32) into (5-34).

$$y(\theta) = 0.985359\theta - 0.142458\theta^3 \qquad (5\text{-}35)$$

196   Function Circuits

I	A(I)	B(I)
1	0.	0.
2	1.570793E+00	1.133697E+00
3	0.	0.
4	-6.456169E-01	-1.380333E-01
5	0.	0.
6	7.102730E-02	4.508397E-03
7	0.	0.
8	7.896531E-02	-7.305285E-05
9	0.	0.
10	-3.987088E-01	-2.244636E-05
11	0.	0.
12	1.028900E+00	-3.052783E-05
13	0.	0.
14	-1.456587E+00	-2.642803E-05
15	0.	0.
16	1.060440E+00	-1.548560E-05
17	0.	0.
18	-3.092132E-01	-4.718219E-06

Fig. 5.11 Generate the coefficients of a series of Tchebysheff polynomials for the sine function defined by 19 sets of x,y points.

Compare Eq. (5-35) with (5-29). Their corresponding coefficients are identical to the fourth significant digits even though Eq. (5-35) is derived by starting with a table of values, while (5-29) is started with the Taylor's series of the sine function.

5.3.2 Thermocouple linearization example[4]

As a second example, consider type E chromel-constantan thermocouples, which are commonly used to

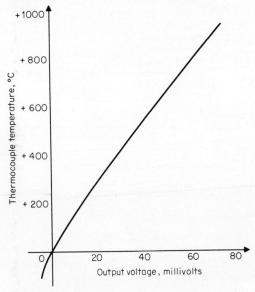

Fig. 5.12 Output voltage and temperature relationship for type E chromel-constantan thermocouple.

measure temperature from -100 to $+900°C$. The output of a thermocouple is in millivolts, and a reference table is sometimes used to convert millivolts to temperature. The output voltage and temperature relationship is plotted in Fig. 5.12. Eleven equally spaced points are taken across the curve, and their values are listed in the first two columns of Table 5.4.

TABLE 5.4 Type E Chromel-Constantan Thermocouple, Approximated by a Fourth-degree Polynomial with Less than 3°C Error.

Thermocouple temperature y (°C)	Thermocouple voltage e (mV)	Normalized $x = 0.0270161e$ $- 0.860057$	Output from approximation polynomial (mV)
-100	-5.18	-1.00000	-97.2
0	0	-0.860057	-2.7
$+100$	$+6.32$	-0.689315	$+99.3$
$+200$	$+13.42$	-0.497501	$+201.3$
$+300$	$+21.04$	-0.291638	$+301.3$
$+400$	$+28.95$	-0.0779409	$+400.1$
$+500$	$+37.01$	0.139809	$+498.5$
$+600$	$+45.10$	0.358369	$+598.5$
$+700$	$+53.14$	0.575579	$+700.0$
$+800$	$+61.08$	0.790086	$+801.1$
$+900$	$+68.85$	1.00000	$+898.8$

The graph is quite linear from $+300$ to $+900°C$ but curved downward from $+300$ to $-100°C$. Now we want to design a black box whose input is the thermocouple millivolts and whose output voltage, also in millivolts, corresponds directly to the thermocouple temperature in degrees Celsius.

First the values of thermocouple millivolts, e, in Table 5.4 are normalized by changing to a new variable x. From Eq. (5-30) and $e_n = +68.85$, $e_1 = -5.18$,

$$x = 0.0270161e - 0.860051 \qquad (5\text{-}36)$$

Use program POWSERS. Enter the values of x and y from Table 5.4 as input data. The program's output data are listed in Fig. 5.13, where the values in the column under A(I) are the coefficients of the tenth-degree polynomial approximating the temperature-voltage curve of Type E thermocouples.

Recall program TCHEB. Enter A(I) from Fig. 5.13 as input data. The coefficients of the series of Tchebysheff polynomials are listed in Fig. 5.14 under B(I). If the first five terms are retained, then

$$y(x) \doteq b_1 + b_2 T_1(x) + b_3 T_2(x) + b_4 T_3(x) + b_5 T_4(x) \qquad (5\text{-}37)$$

N	X(I)	Y(I)
1	-1.000000E+00	-1.000000E+02
2	-8.600570E-01	0.
3	-6.893150E-01	1.000000E+02
4	-4.975010E-01	2.000000E+02
5	-2.916380E-01	3.000000E+02
6	-7.794090E-02	4.000000E+02
7	1.398090E-01	5.000000E+02
8	3.583690E-01	6.000000E+02
9	5.755790E-01	7.000000E+02
10	7.900860E-01	8.000000E+02
11	1.000000E+00	9.000000E+02

N	A(I)
1	4.359354E+02
2	4.597724E+02
3	-1.464493E+01
4	2.463566E+01
5	9.097553E+00
6	5.882029E+00
7	-8.055010E+01
8	2.998474E+00
9	1.033681E+02
10	6.711417E+00
11	-5.320599E+01

Fig. 5.13 A(I) are the coefficients of the polynomial approximating the temperature-voltage relationship of type E thermocouples.

where $b_1 = 422.0237$
$b_2 = 486.8685$
$b_3 = -17.13077$
$b_4 = 11.18311$
$b_5 = -3.824332$

From Table 5.1, substitute $T_1(x)$, $T_2(x)$, $T_3(x)$, and $T_4(x)$ into Eq. (5-37) and simplify.

$$y(x) \doteq 435.330 + 453.319x - 3.66686x^2 + 44.7324x^3 - 30.5947x^4 \quad (5\text{-}38)$$

I	A(I)	B(I)
1	4.359354E+02	4.220237E+02
2	4.597724E+02	4.868685E+02
3	-1.464493E+01	-1.713077E+01
4	2.463566E+01	1.118311E+01
5	9.097553E+00	-3.824332E+00
6	5.882030E+00	1.639378E+00
7	-8.055010E+01	-7.329921E-01
8	2.998474E+00	2.827994E-01
9	1.033681E+02	-2.316162E-01
10	6.711417E+00	2.621647E-02
11	-5.320599E+01	-1.039179E-01

Fig. 5.14 Coefficients of the series of Tchebysheff polynomials approximating the temperature-voltage curve of type E thermocouples.

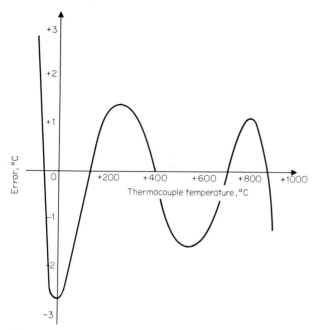

Fig. 5.15 The approximation of the type E thermocouple's temperature-voltage curve has less than 3°C error.

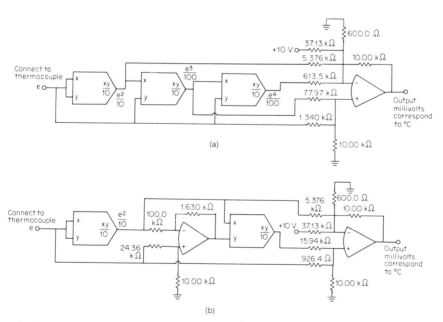

Fig. 5.16 (a) Output voltage in millivolts corresponds directly to the temperature of Type E thermocouples in degrees Celsius. (b) The same transfer function using two multipliers and two operational amplifiers.

Change x back to the original variable e by substituting Eq. (5-36) into (5-38).

$$y(e) \doteq -2.69309 + 17.2024e - 0.186021e^2 + 0.00295744e^3 \\ - 0.0000163000e^4 \quad (5\text{-}39)$$

The values of this approximation are listed in the fourth column in Table 5.4, and a plot of the error curve is shown in Fig. 5.15. The maximum error is less than 3°C. Figure 5.16a shows a simulation employing three multipliers and one operational amplifier. The same approximation can be implemented by using two multipliers and two operational amplifiers, as shown in Fig. 5.16b, if Eq. (5-39) is rewritten in the form

$$y(e) = -2.69309 + 17.2024e - 0.186021e^2 + e^2(0.00295744e \\ - 0.0000163e^2)$$

A buffer amplifier may be needed between the thermocouple and the input terminal to avoid loading errors.

5.3.3 Inverse function generators[6,8]

The inverse function $x = f^{-1}(y)$, where $y = f(x)$ is given, can be generated by putting the $f(x)$ module in the feedback loop of an operational amplifier, as shown in Fig. 5.17. The $f(x)$ module should always be monotonic and single-valued; for example, $\sin^{-1} y$ should be limited to two quadrants only, from $-\pi/2$ to $+\pi/2$. Gain and phase margin should be investigated to avoid oscillation.

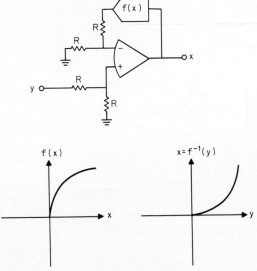

Fig. 5.17 Inverse function $x = f^{-1}(y)$ generated by putting $f(x)$ module in the feedback loop of an operational amplifier.

5.4 Approximation Using Noninteger Exponent[5,6]

The theory of operation of the multifunction converter, $e_o = v_y (v_z/v_x)^m$, can be found in Chapter 3. We will take advantage of its capability to raise a variable to a noninteger exponent. Three of the most commonly used functions, sine, cosine, and arctangent, will be approximated with noninteger exponents to achieve higher accuracy. Actual circuit implementations and adjustment procedures will also be discussed.

5.4.1 Sine-function generator

A theoretical error of less than 0.25 percent in the first quadrant is made possible by the following approximation with only two terms:

$$\sin x \doteq x - \frac{x^{2.827}}{6.28} \qquad (5\text{-}40)$$

A plot of the error curve is given in Fig. 5.18. Most function modules in the market are optimized to swing 10 V with ± 15 V power supplies. To

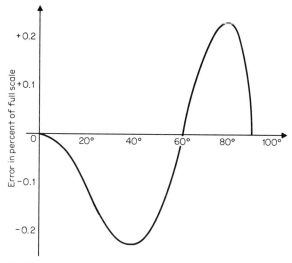

Fig. 5.18 The sine-function approximation with two terms and noninteger exponent has less than 0.25 percent error in the first quadrant.

actually implement the sine function, component tolerance effects will be minimized if it is scaled to utilize the whole dynamic range of 10 V, i.e., if 90° corresponds to +10 V, as shown in Fig. 5.19a. The transfer function becomes

$$e_o = 10 \sin 9e_i \qquad (5\text{-}41)$$

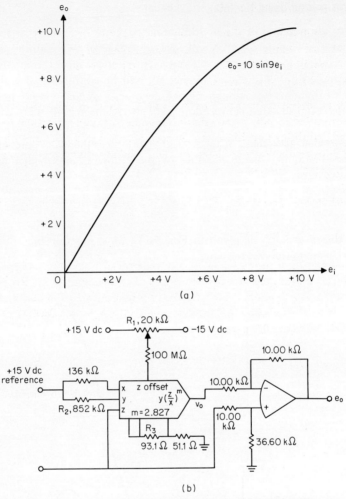

Fig. 5.19 (a) and (b). One-quadrant sine function, scaled such that 90° corresponds to +10 V in (a), can be implemented as in (b).

Changing variables in Eq. (5-40) and substituting into Eq. (5-41) yields

$$e_o \doteq 1.571 e_i - 1.592 \left(\frac{e_i}{6.366}\right)^{2.827}$$

Figure 5.19b shows the circuit configuration using one multifunction converter in conjunction with an operational amplifier. The circuit's overall accuracy is very sensitive to the multifunction converter's z input offset voltage, and R_1 is provided to trim it out. The following adjustment procedures should be followed to compensate for component tolerances. It has been experimentally determined that the generator has less than ±50 mV actual measured error.

Adjustment procedures:

1. Adjust R_1 such that v_o is less than 1 mV when $e_i = 0$.
2. Adjust R_2 such that $v_o = 0.8045$ V dc when $e_i = 5.000$ V dc.
3. Adjust R_3 such that $v_o = 5.709$ V dc when $e_i = 10.00$ V dc.
4. Repeat steps 2 and 3 as necessary.

The reference voltage is shown to be +15 V dc since in most cases the ±15 V dc supplied to power the circuit has sufficient time- and temperature-related stabilities to achieve the desired accuracy. If the particular power supplies available do not have the necessary stability, an additional +15 V dc precision reference may be required.

5.4.2 Cosine-function generator

The Taylor's series expansion of $\cos x$ is

$$\cos x = 1 - \frac{x^2}{2!} + \frac{x^4}{4!} - \frac{x^6}{6!} + \cdots$$

Approximation with three terms will have an error of 2 percent, and two multipliers or multifunction converters and one operational amplifier will be required to implement it. By using a noninteger exponent and one linear term, it is possible to approximate, as in Eq. (5-42), the cosine function with less than 0.8 percent error in the first quadrant.

$$\cos x \doteq 1 + 0.235x - \frac{x^{1.504}}{1.445} \qquad (5\text{-}42)$$

The error curve is plotted in Fig. 5.20. The actual transfer function is optimized if it is scaled such that

$$e_o = 10 \cos 9e_i \qquad (5\text{-}43)$$

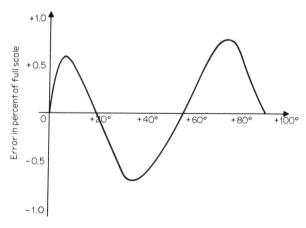

Fig. 5.20 Error curve to approximate $\cos x$ by $1 + 0.235x - \dfrac{x^{1.504}}{1.455}$.

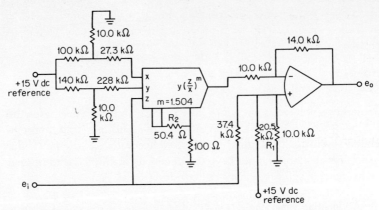

Fig. 5.21 Cosine-function generator with less than 0.8 percent measured error.

Substitute Eq. (5-42) into (5-43):

$$e_o \doteq 10 + 0.3652 e_i - 0.4276 e_i^{1.504} \tag{5-44}$$

Figure 5.21 presents the circuit implementing Eq. (5-44). The circuit should be adjusted as follows:

1. Adjust R_1 such that $e_o = +10.00$ V dc when $e_i = 0$.
2. Adjust R_2 such that $e_o = 0$ when $e_i = +10.00$ V dc.

Actual experiment indicates that the overall error, after adjustments, approaches the theoretical error of ± 80 mV.

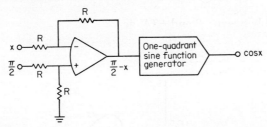

Fig. 5.22 Implementing $\cos x$ in the first quadrant by $\sin\left(\dfrac{\pi}{2} - x\right)$.

By using one more operational amplifier, $\cos x$ can be simulated as $\sin\left(\dfrac{\pi}{2} - x\right)$ and 0.25 percent theoretical error is achievable, as shown in Fig. 5.22.

5.4.3 Arctangent and vector magnitude The arctangent of a ratio and the square root of the sum of squares are two companion functions widely used to convert from rectangular coordinates to polar coordinates.

Tan⁻¹ z is one of the most difficult functions to approximate because z changes from zero to infinity as tan⁻¹ z ranges from 0 to 90°. Equation (5-45) utilizes noninteger exponents to approximate tan⁻¹ z and has less than 0.75 percent theoretical error over the whole dynamic range.

$$\tan^{-1} z \doteq \frac{\pi}{2} \frac{z^{1.2125}}{1 + z^{1.2125}} \qquad (5\text{-}45)$$

A plot of the error curve can be found in Fig. 5.23. Substituting $z = e_y/e_x$ into Eq. (5-45) to get the arctangent of a ratio and converting to degrees,

$$\tan^{-1} \frac{e_y}{e_x} = \frac{(e_y/e_x)^{1.2125}}{1 + (e_y/e_x)^{1.2125}} (90°)$$

The circuit implementing $\tan^{-1}(e_y/e_x)$ is shown in Fig. 5.24. The 25 kΩ potentiometer should be adjusted such that with $e_x = e_y = +10.00$ V dc, $e_o = +4.500$ V dc ± 1 mV. Actual measured error of the arctangent generator depends on the ranges of e_x and e_y. For e_x and e_y between 10 and 2 V, the overall error is typically ±55 mV; for e_x and e_y between 2 and 0.1 V, the error is typically ±65 mV; and for e_x and e_y between 0.1 and 0.03 V, the error is typically ±340 mV.

The companion circuit to $\tan^{-1}(e_y/e_x)$ is $\sqrt{e_x^2 + e_y^2}$; this can be built with one multifunction converter and two operational amplifiers, as shown in Fig. 5.25. Three diodes are used to prevent the circuit from latching up to the negative supply voltage. All resistors should be matched to 10.00 kΩ ± 0.02%, and the 25 kΩ potentiometer should be adjusted so that when $e_y = 0$ and $e_x = +10.00$ V dc, $e_o = +10.00$ V dc. An overall error of less than ±7 mV has been achieved when e_x is operated between 0 and +6 V dc and e_y is operated between −3 and +3 V dc.

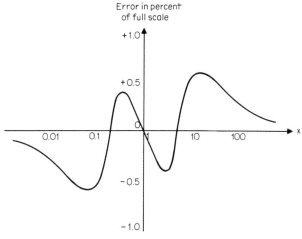

Fig. 5.23 Error curve of approximation of $\tan^{-1} x$ by $\dfrac{\pi}{2} \dfrac{z^{1.2125}}{1 + z^{1.2125}}$

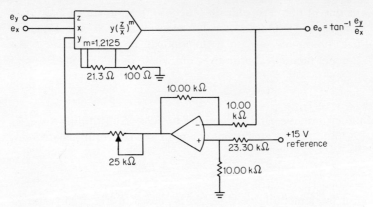

Fig. 5.24 Arctangent generator, $e_o = \tan^{-1} \frac{e_y}{e_x}$, with less than 0.75 percent theoretical error.

The vector magnitude circuit can easily be expanded to include n terms. In equation form,

$$e_o = \sqrt{v_1^2 + v_2^2 + v_3^2 + \cdots + v_n^2} \qquad (5\text{--}46)$$

Equation (5–46) can be rewritten as

$$e_o = v_1 + \frac{v_2^2}{e_o + v_1} + \frac{v_3^2}{e_o + v_1} + \cdots + \frac{v_n^2}{e_o + v_1} \qquad (5\text{--}47)$$

The circuit configuration implementing Eq. (5–47) is shown in Fig. 5.26. It is obvious that for n terms it requires two operational amplifiers and

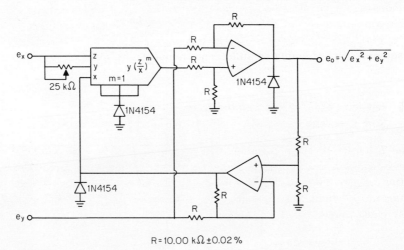

Fig. 5.25 Vector magnitude circuit, $e_o = \sqrt{e_x^2 + e_y^2}$.

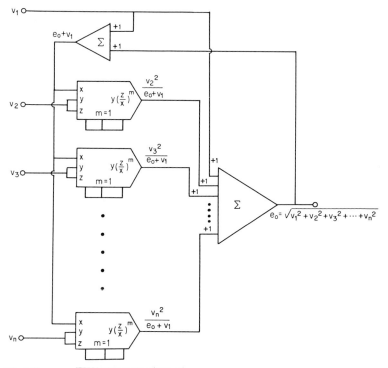

Fig. 5.26 $e_o = \sqrt{v_1^2 + v_2^2 + v_3^2 + \cdots + v_n^2}$ requires two operational amplifiers and $(n-1)$ multifunction converters.

$(n-1)$ multifunction converters or function modules performing xy/z. Diodes to prevent latch-up like those used in Fig. 5.25 may also be needed. This approach provides much better accuracy over a wide dynamic range of e_o than the straightforward method of implementing Eq. (5-46) using n squarers, one summing amplifier, and then a square rooter.

5.5 Implicit Feedback and Four-Quadrant Sine Generator[6]

The Tchebysheff polynomials method presented in Secs. 5.2.1 and 5.2.2 can be applied to approximate a sine function in all four quadrants. Two terms are generally not enough to achieve reasonable accuracies. If three terms are retained, the approximation, as given in Eq. (5-48), will yield less than 0.75 percent error from $-180°$ to $+180°$.

$$e_o = \sin e_i \\ \doteq 0.98402 e_i - 0.15328 e_i^3 + 0.0054523 e_i^5 \qquad (5\text{-}48)$$

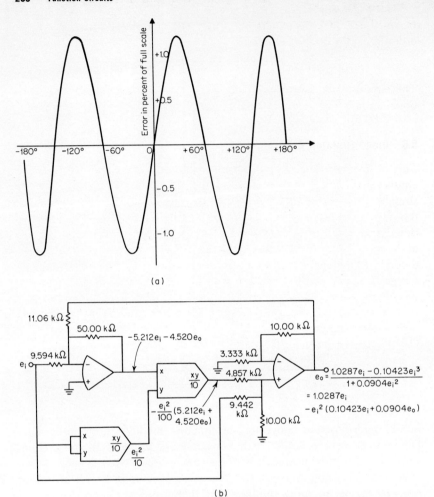

Fig. 5.27 (a) Error curve of approximation of four-quadrant sine function with implicit feedback. (b) The circuit implemented with two multipliers and two operational amplifiers.

The circuit is the same as in Fig. 5.2 except that the resistor values should be recalculated to agree with the coefficients in Eq. (5–48).

However, if implicit feedback is utilized, the four-quadrant sine function can be approximated by

$$e_o = \sin e_i \\ \doteq \frac{1.0287 e_i - 0.10423 e_i^3}{1 + 0.0904 e_i^2} \quad (5\text{–}49)$$

The above equation can be obtained by using numerical methods with a computer. Approximation by Eq. (5–49) has less than 1.25 percent error in four quadrants. A plot of the error curve is given in Fig. 5.27a. In order to model Eq. (5–49), rewrite it as

$$e_o = 1.0287e_i - e_i^2(0.10423e_i + 0.0904e_o)$$

The circuit configuration, which requires two multipliers and two operational amplifiers, is given in Fig. 5.27b.

5.6 Diode Function Generators[2,8]

This chapter on function generators would not be complete without including diode function generators (DFGs). DFGs are designed to provide straight line segment approximations to arbitrary continuous-voltage transfer functions, as shown in Fig. 5.28. The accuracy of the approximation depends on the number of segments used and the shape or complexity of the transfer function. Line segments are generated by summing biased diodes with an operational amplifier. Each diode has zero response until a break point is reached, then it contributes linearly. Since the forward voltage drop across a silicon diode is proportional to ambient temperature in degrees Kelvin (Chapter 2), temperature-compensation circuits are normally required in practical DFGs.

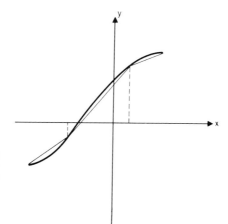

Fig. 5.28 Diode function generator provides straight line segment approximations to arbitrary voltage transfer functions.

5.6.1 Temperature-compensated diode function generator
A general-purpose temperature-compensated DFG is shown in Fig. 5.29a, with the x offset control circuit shown in Fig. 5.29b. Ten fixed break points are equally spaced from -10 to $+10$ V. Eleven slope controls vary the slope of each segment plus or minus 2 V/V. The x and y axes may be shifted by means of x and y offset controls. An input buffer gain control permits reduction of the effective break-point spacing for input signals of less than 10 V peak

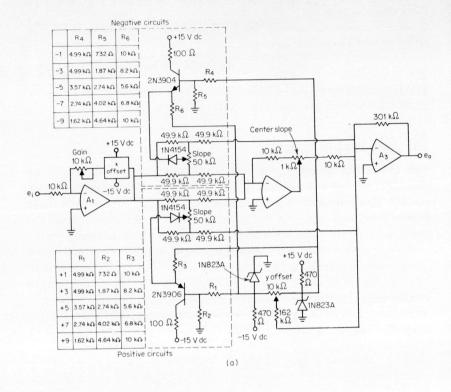

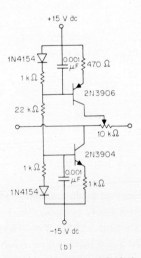

Fig. 5.29 (a) General-purpose temperature-compensated diode function generator with 10 break points. (b) x offset adjustment circuit.

amplitude. The gain control and x offset control may be employed to concentrate all the break points to the right or left of the y axis and provide an 11-segment output function for unipolar input signals. Special-purpose DFGs may be derived from this general-purpose one by replacing potentiometers with fixed resistors and by reducing the number of break points as required.

5.6.2 Theory of operation

Input signals are applied through the 10 kΩ summing resistor of an inverting input buffer amplifier A_1, whose feedback network includes the circuitry for the x offset control, which is continuously adjustable from -10 to $+10$ V, and also a gain adjustment potentiometer, which is continuously adjustable from 1 to 2. The signal then passes to 10 slope control networks in parallel, five of which are for positive break points and five for negative break points. Each network contains one 50 kΩ potentiometer for slope control and has two outputs, one of which is inverted in the inverting amplifier A_2. The difference between the inverted and noninverted slope network outputs is amplified by the output buffer amplifier A_3. The difference is zero until the break-point voltage is reached. The silicon diode in each slope control network is temperature-compensated by the forward-biased base-emitter junction of the emitter-follower transistor. When the break-point voltage is exceeded, a difference voltage, whose magnitude and polarity are determined by the position of the slope control, is amplified and applied to the output terminal. The center slope control adjusts the gain of the inverting amplifier A_2 to set the slope of the center segment between the -1 and $+1$ break points. Each slope control covers a range from -2 to $+2$ V/V. A variable bias voltage is applied to the output buffer amplifier A_3 to provide a y offset adjustable between -10 and $+10$ V.

5.6.3 Function setup

Adjustment of the controls may be carried out with the aid of an oscilloscope or an x-y plotter. A 20 V p-p triangular or sinusoidal signal is applied to the input of the DFG and to the horizontal input of an oscilloscope or the x axis of a plotter. The output is connected to the vertical input of the oscilloscope or y axis of the plotter.

Fitting a given function, which may be traced on the scope graticule or on a sheet of paper on the x-y plotter, is carried out by adjusting the incremental slope controls and the x and y offset controls. The x and y offset controls are normally adjusted first. The x offset is monitored at the output of amplifier A_1 with the input e_i grounded. The y offset is monitored at e_o with the input grounded and the x offset set at zero or all the slope controls set at zero slope. Next the center slope and the other slope controls are set, beginning with the center slope and working outward one segment at a time. The function may be more precisely set using dc inputs and a precision dc voltmeter.

The effective break-point spacing may be reduced to 1 V when the input buffer amplifier gain control is turned to its maximum setting. When the control is set for maximum gain, an 11-segment output function can be generated from input signals that have a 10 V p-p amplitude. The x offset control can then be used to shift the break points either to the left or to the right of the y axis for signals which are not symmetrical about zero.

REFERENCES

1. I. S. Sokolnikoff and R. M. Redheffer, *Mathematics of Physics and Modern Engineering*, 2d ed., McGraw-Hill Book Company, New York, 1966.
2. G. A. Korn and T. M. Korn, *Electronic Analog and Hybrid Computers*, 2d ed., McGraw-Hill Book Company, New York, 1972.
3. J. Vlach, *Computerized Approximation and Synthesis of Linear Networks*, John Wiley & Sons, Inc., New York, 1969.
4. *Temperature Measurement Thermocouples*, Instrument Society of America, Pittsburgh, Pa., 1964.
5. D. H. Sheingold, Approximate Analog Function with a Low-Cost Multiplier/Divider, *EDN*, Feb. 5, 1973.
6. D. H. Sheingold, *Nonlinear Circuits Handbook*, Analog Devices, Inc., Norwood, Mass., 1974.
7. *USC-VI Fortran Reference Manual*, United Computing Systems, Inc., Kansas City, Mo., October 1972.
8. G. E. Tobey, J. G. Graeme, and L. P. Huelsman, *Operational Amplifiers: Design and Applications*, McGraw-Hill Book Company, New York, 1971.

6
SIMPLIFIED DESIGN OF ACTIVE FILTERS

An active filter is a collection of resistors, capacitors, and active elements (mostly operational amplifiers). Viewed in another sense, it is a circuit without inductors. Active filters have many advantages over *RLC* filters in the audio and subaudio frequency ranges. First of all, the inductor is a relatively large and heavy element. Second, inductors generally have more dissipation associated with them than capacitors of similar size do. In other words, commercially available inductors are not nearly as "ideal" as commercially available capacitors usually are. If you have tried to use network synthesis techniques, you have probably discovered that the dissipation (or resistance) associated with inductors can cause considerable difficulty. For these reasons and a few others such as nonlinearity, saturation, and cost, more and more interest is being shown in design of filters without inductors, namely, active *RC* filters.

Like passive *RLC* filters, active *RC* filters can have natural frequencies anywhere in the left half of the complex frequency (or s) plane. They can function as oscillators, i.e., they can have natural frequencies on the $j\omega$ axis. They can provide transformation ratios just as coupled coils of a transformer do; however, they cannot provide the isolation. They can provide perfect coupling and thus realize "ideal" transformers, which actual coupled coils cannot do. They can even gyrate microfarads of capacitance into hundreds of henries of inductance.

The tremendous capability of active RC filters comes mainly from the active element (operational amplifier). The passive RC elements, by themselves, can produce natural frequencies only on the negative real axis of the complex frequency plane. Right half plane natural frequencies, of course, are not useful because they signify unstable network behavior. We will consider the usable active RC natural frequencies as being those in the left half plane or on the $j\omega$ axis.

6.1 Synthesis of Active RC Networks

Many new active filter circuits have been developed in recent years. Some of these circuits have been of great theoretical interest, but of little practical value. Others, however, are extremely practical and have great potential for everyday applications. The goal of this section has been to screen the large volume of literature on this subject and present only those techniques which are of definite practical value to the working engineer.

6.1.1 Second-order low-pass filters For a second-order low-pass filter with complex conjugate poles, the voltage transfer function is of the form

$$\frac{E_o(s)}{E_i(s)} = \frac{K\omega_o^2}{s^2 + s\omega_o/Q + \omega_o^2} \qquad (6\text{-}1)$$

where K is the dc gain factor, ω_o is the cutoff frequency, and Q is a damping constant determining the selectivity of the frequency response around ω_o. Two widely used low-pass circuits are the voltage-controlled voltage source (VCVS) and the infinite-gain multiple-feedback (MF). Their respective circuits are shown in Fig. 6.1a and b.

VCVS low-pass active filter. In terms of the elements shown in Fig. 6.1a, we may write the voltage transfer function as

$$\frac{E_o(s)}{E_i(s)} = \frac{m/(R_1R_2C_1C_2)}{s^2 + s\,[1/R_1C_1 + 1/R_2C_1 - (m-1)/R_2C_2] + 1/R_1R_2C_1C_2} \qquad (6\text{-}2)$$

Comparing Eqs. (6–1) and (6–2), we can write the following three simultaneous equations:

$$K\omega_o^2 = \frac{m}{R_1R_2C_1C_2} \qquad (6\text{-}3)$$

$$\frac{\omega_o}{Q} = \frac{1}{R_1C_1} + \frac{1}{R_2C_1} - \frac{m-1}{R_2C_2} \qquad (6\text{-}4)$$

$$\omega_o^2 = \frac{1}{R_1R_2C_1C_2} \qquad (6\text{-}5)$$

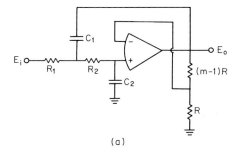

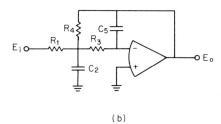

Fig. 6.1 (a) Voltage-controlled voltage source low-pass filter. (b) Infinite-gain multiple-feedback low-pass filter.

To design a practical filter circuit with given ω_o, Q, and K, always make $C_1 = C_2 = C$ and $R_1 = R_2 = R$. With a convenient value chosen for C, the following design formulas result from Eqs. (6–3), (6–4), and (6–5):

$$m = 3 - \frac{1}{Q}$$

and

$$R = \frac{1}{\omega_o C}$$

Note that with this design, $K = m$. R_1 can be replaced by a resistor divider to lower the value of K so long as the equivalent resistance of the divider network equals R.

MF low-pass active filter. The voltage transfer function for the MF circuit in Fig. 6.1b is

$$\frac{E_o(s)}{E_i(s)} = \frac{-1/R_1 R_3 C_2 C_5}{s^2 + s\,(1/C_2)(1/R_1 + 1/R_3 + 1/R_4) + 1/R_3 R_4 C_2 C_5} \qquad (6\text{--}6)$$

The specific solutions for the element values in terms of given parameters ω_o, Q, and K may be found by equating corresponding coefficients in Eqs. (6–1) and (6–6). Such a process leads to a set of three simultaneous nonlinear equations. Solving the set of equations results in the following design formulas:

$$R_1 = \frac{1/Q \pm \sqrt{1/Q^2 - 4\,(|K| + 1)(C_5/C_2)}}{2|K|\,\omega_o C_5}$$

$$R_4 = |K| R_1$$

and

$$R_3 = \frac{1}{\omega_0^2 R_4 C_2 C_5}$$

6.1.2 Second-order high-pass filters High-pass filters provide sharp attenuation below the cutoff frequency and a flat gain characteristic above. They are often used to block the dc component of a signal while passing the alternating component of the signal with low distortion. Active high-pass filters with complex conjugate poles are an order of magnitude better than simple capacitor coupling for this type of application. Another application for high-pass filters arises where there is a need to detect a small high-frequency signal that is superimposed on a large low-frequency signal. The high-pass filter will reject the low-frequency signal while passing the high-frequency signal.

The standard voltage transfer function for a second-order high-pass filter with complex conjugate poles is of the form

$$\frac{E_o(s)}{E_i(s)} = \frac{Ks^2}{s^2 + s\omega_0/Q + \omega_0^2} \qquad (6\text{--}7)$$

where K is the gain factor in the passband, ω_0 is the cutoff frequency, and Q is the damping constant. There are as many high-pass filter circuits around as low-pass circuits, e.g., the infinite-gain single-feedback technique, the infinite-gain multiple-feedback technique, a technique using negative immittance converters, a technique using controlled sources, and the state-variable technique. For general-purpose low-cost applications, the voltage-controlled voltage source (VCVS) technique is relatively simple and easy to use. The VCVS high-pass circuit is given in Fig. 6.2. The transfer function, in terms of circuit elements, is

$$\frac{E_o(s)}{E_i(s)} = \frac{s^2 m}{s^2 + s\,[1/R_2 C_2 + 1/R_2 C_1 - (m-1)/R_1 C_1] + 1/R_1 R_2 C_1 C_2} \qquad (6\text{--}8)$$

By equating the corresponding coefficients in Eqs. (6-7) and (6-8), we have the three simultaneous nonlinear equations

$$K = m$$

$$\omega_0^2 = \frac{1}{R_1 R_2 C_1 C_2}$$

and

$$\frac{\omega_0}{Q} = \frac{1}{R_2 C_2} + \frac{1}{R_2 C_1} - \frac{m-1}{R_1 C_1}$$

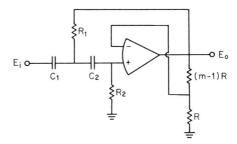

Fig. 6.2 Voltage-controlled voltage source high-pass active filter.

As in the VCVS low-pass case, always make $C_1 = C_2 = C$ and $R_1 = R_2 = R$. Solving the set of nonlinear equations, we have the design formulas:

$$m = 3 - \frac{1}{Q}$$

and

$$R = \frac{1}{\omega_o C}$$

It should be noted that although the above design formulas have been found to give excellent experimental results, they are not unique, i.e., other sets of solutions also exist.

6.1.3 Active bandpass filters

Bandpass filters are widely used to remove noise or adjacent-band interference from a signal consisting of a single frequency or a narrow band of frequencies. Most bandpass filter applications require that the Q factor be high, i.e., that the poles be close to the $j\omega$ axis. This makes the design more difficult, since the stability of the filter circuit becomes critical for high Q factors. The majority of bandpass filters are symmetrical. As shown in Fig. 6.3, they are geometric-mean symmetrical about a center frequency. The following two equations hold for any symmetrical bandpass filter:

$$f_o^2 = f_1 f_2$$

and

$$Q = \frac{f_o}{f_2 - f_1}$$

where f_o is the center frequency, f_1 and f_2 are the lower and upper -3 dB frequencies, respectively, and Q is the damping constant. The attenuation in decibels at xf_o is always the same as that at f_o/x.

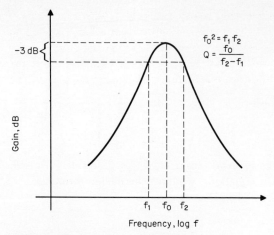

Fig. 6.3 Symmetrical bandpass filter characteristics.

The voltage transfer function of a second-order bandpass filter is of the form

$$\frac{E_o(s)}{E_i(s)} = \frac{sK\omega_0/Q}{s^2 + s\omega_0/Q + \omega_0^2} \quad (6\text{-}9)$$

where $\omega_0 = 2\pi f_0$ and K is the gain constant at f_0. Note that the phase shift of Eq. (6-9) at $\omega = \omega_0$ is zero. This unique characteristic provides a power-

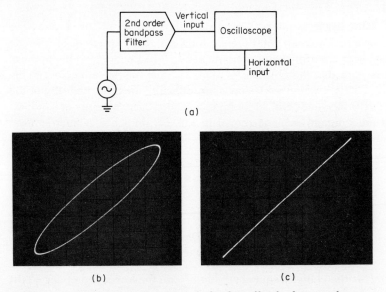

Fig. 6.4 (a) Tuning the center frequency of a bandpass filter by detecting the zero phase-shift point. (b) The input and output are out of phase, with the oscilloscope trace showing an ellipse. (c) They are in phase when the ellipse collapses and forms a straight line.

ful method of tuning the center frequency of bandpass filters. As shown in Fig. 6.4a, simply connect the input of the filter to the horizontal input of the oscilloscope and connect the filter output to the vertical input of the oscilloscope. An ellipse will appear on the scope (Fig. 6.4b) when the input and output of the filter are out of phase, i.e., when $\omega \neq \omega_0$. The

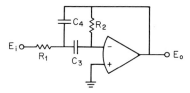

Fig. 6.5 Infinite-gain multiple-feedback bandpass filter.

ellipse will collapse and become a straight line (Fig. 6.4c) when they are in phase and $\omega = \omega_0$.

To build high-Q bandpass filters, the state-variable technique should be used, since it has low sensitivity to circuit elements and also is easy to tune. For general-purpose low-cost applications, the infinite-gain multiple-feedback circuit requires the minimum number of components and gives satisfactory results for low Q factors ($Q < 10$). A detailed discussion of the state-variable technique will be presented in Sec. 6.1.5. Figure 6.5 shows the circuit diagram of the multiple-feedback bandpass filter. The voltage transfer function may be written in terms of circuit elements as

$$\frac{E_o(s)}{E_i(s)} = \frac{-s/R_1 C_4}{s^2 + s(1/C_3 + 1/C_4)/R_2 + 1/R_1 R_2 C_3 C_4} \quad (6\text{--}10)$$

By equating the corresponding coefficients of Eqs. (6–9) and (6–10) and solving the three simultaneous nonlinear equations, the design formulas for R_1, R_2, and K become

$$R_1 = \frac{1}{\omega_0 Q (C_3 + C_4)}$$

$$R_2 = \frac{Q}{\omega_0} \left(\frac{1}{C_3} + \frac{1}{C_4} \right)$$

and

$$|K| = Q^2 \left(1 + \frac{C_3}{C_4} \right)$$

If we make $C_3 = C_4 = C$, the design formulas are simplified to

$$R_1 = \frac{1}{2\omega_0 QC}$$

$$R_2 = \frac{2Q}{\omega_o C}$$

and

$$|K| = 2Q^2$$

6.1.4 Active band-reject filters Band-reject filters place an attenuation "notch" in the frequency response at the center frequency f_o. As with symmetrical bandpass filters, the Q factor which determines the selectivity of the notch is given by

$$Q = \frac{f_o}{f_2 - f_1}$$

and

$$f_o^2 = f_1 f_2$$

where f_1 and f_2 are the -3 dB frequencies. Notch filters are usually used to eliminate specific interfering frequencies. They are often cascaded with other filters to form special response characteristics. The standard transfer function for a second-order symmetrical notch filter is of the form

$$\frac{E_o(s)}{E_i(s)} = \frac{K(s^2 + \omega_o^2)}{s^2 + s\omega_o/Q + \omega_o^2} \qquad (6\text{--}11)$$

where $\omega_o = 2\pi f_o$ and K is the gain constant in the passband, i.e., at frequencies much lower or higher than f_o.

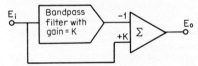

Fig. 6.6 Build a band-reject filter by subtracting the output from the input of a bandpass filter.

Convert a bandpass filter to a notch filter. A notch filter can be built by subtracting the input from the output of a bandpass filter, as illustrated in Fig. 6.6. The theory is based on the mathematical identity

$$\frac{K(s^2 + \omega_o^2)}{s^2 + s\omega_o/Q + \omega_o^2} = K - \frac{Ks\omega_o/Q}{s^2 + s\omega_o/Q + \omega_o^2} \qquad (6\text{--}12)$$

The expression on the left-hand side of Eq. (6–12) corresponds to the transfer function of a second-order band-reject filter, while the second term on the right-hand side of Eq. (6–12) corresponds to the transfer

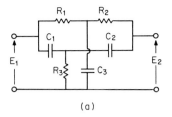

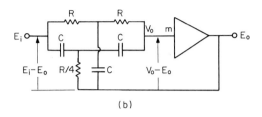

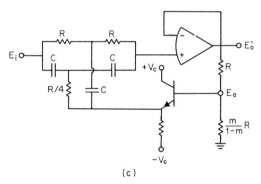

Fig. 6.7 (a) Twin-T network may be used as a passive notch filter with passive poles and zeros. (b) High-Q notch filter with complex conjugate poles requires an additional voltage amplifier with $(1 - \sqrt{2}/3) < m < 1$. (c) Complete circuit diagram of an active twin-T notch filter.

function of a bandpass filter. This active notch filter requires at least two operational amplifiers, one for the bandpass portion and the other for the output difference amplifier.

Twin-T band-reject filter. The familiar twin-T RC network is shown in Fig. 6.7a. If the right combination of RC values is chosen, the network will work as a passive notch filter with simple poles and zeros. One such combination is $C_1 = C_2 = C_3 = C$, $R_1 = R_2 = R$, and $R_3 = R/4$. The voltage transfer function of the network then becomes

$$\frac{E_2(s)}{E_1(s)} = \frac{s^2 + 2/(RC)^2}{s^2 + s6/RC + 2/(RC)^2} \qquad (6\text{--}13)$$

By equating the corresponding coefficients in Eqs. (6-11) and (6-13) and simplifying, it can be found that

$$\omega_o = \frac{\sqrt{2}}{RC}$$

and

$$Q = \frac{\sqrt{2}}{6}$$

By connecting the twin-T RC network in the feedback loop of a voltage amplifier, as illustrated in Fig. 6.7b, high-Q notches with complex conjugate poles are achievable. Let us derive the transfer function for this active twin-T notch filter. The transfer function for the passive twin-T portion is

$$\frac{V_o - E_o}{E_i - E_o} = \frac{s^2 + \omega_T^2}{s^2 + s\omega_T/Q_{PT} + \omega_T^2} \quad (6\text{-}14)$$

where $\omega_T = \sqrt{2}/RC$ and $Q_{PT} = \sqrt{2}/6$. For the output voltage amplifier,

$$E_o = mV_o \quad (6\text{-}15)$$

Eliminating V_o from Eqs. (6-14) and (6-15) and simplifying, the resulting transfer function of E_o/E_i becomes

$$\frac{E_o(s)}{E_i(s)} = \frac{m(s^2 + \omega_T^2)}{s^2 + s\omega_T/Q_T + \omega_T^2} \quad (6\text{-}16)$$

where $Q_T = Q_{PT}/(1 - m)$. Note that m must be less than unity or the poles of Eq. (6-16) will move to the right half s plane. To have complex conjugate poles, $Q_T > 0.5$, that is, $m > (1 - \sqrt{2}/3)$. The voltage amplifier of Fig. 6.7b can be built with one operational amplifier, a voltage divider, and an emitter follower, as illustrated in Fig. 6.7c. This output stage has the added advantage that $E'_o = E_o/m$, and hence the filter's voltage transfer function becomes

$$\frac{E'_o(s)}{E_i(s)} = \frac{s^2 + \omega_T^2}{s^2 + s\omega_T/Q_T + \omega_T^2}$$

6.1.5 The universal active filter

The universal active filter uses the state-variable technique to produce a basic second-order filter transfer function. Three separate outputs provide low-pass, high-pass, and bandpass transfer functions. A band-reject transfer function may be realized simply by summing the high-pass and low-pass outputs. Because of its versatility, this is given the name *universal active filter* (UAF). The UAF is especially suitable for high-Q applications because of its low gain and Q sensitivities.

It provides the user with easy control of the Q factor, resonant frequency, and gain. Any complex filter response can be obtained by cascading units. The UAF is an ideal basic filter building block that can be stocked in quantity to be used whenever the requirement for a filter arises.

The state-variable technique uses two operational-amplifier integrators and a summing amplifier, as shown in Fig. 6.8. The voltage transfer functions between E_L/E_i, E_H/E_i, and E_B/E_i are, respectively,

$$\frac{E_L(s)}{E_i(s)} = \frac{K_L\omega_0^2}{s^2 + s\omega_0/Q + \omega_0^2} \quad \text{low-pass}$$

$$\frac{E_H(s)}{E_i(s)} = \frac{K_H s^2}{s^2 + s\omega_0/Q + \omega_0^2} \quad \text{high-pass}$$

and

$$\frac{E_B(s)}{E_i(s)} = \frac{K_B s\omega_0/Q}{s^2 + s\omega_0/Q + \omega_0^2} \quad \text{bandpass}$$

where $\omega_0 = \sqrt{\dfrac{K_3}{R_1 R_2 C_1 C_2}}$

$$Q = \frac{1+K_4}{1+K_3}\sqrt{\frac{K_3 R_1 C_1}{R_2 C_2}}$$

$$K_L = \frac{K_4(1+K_3)}{K_3(1+K_4)}$$

$$K_H = \frac{K_4(1+K_3)}{1+K_4}$$

$$K_B = -K_4$$

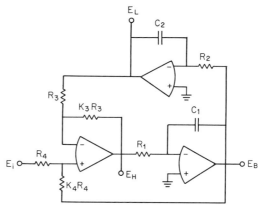

Fig. 6.8 The universal active filter uses the state-variable technique, requiring two operational amplifier integrators and a summing amplifier.

To obtain band-reject characteristics, an additional operational amplifier is needed to sum the low-pass and high-pass outputs and thus form a pair of $j\omega$-axis zeros:

$$\frac{E_R(s)}{E_i(s)} = \frac{K_R(s^2 + \omega_0^2)}{s^2 + s\omega_0/Q + \omega_0^2}$$

where $K_R = K_L = K_H$. The circuit connections are illustrated in Fig. 6.9.

So far, in the mathematical derivations, we have assumed that the operational amplifiers are ideal. Two interesting phenomena occur, especially at high Q's and high frequencies, when you actually build a filter. First, the circuit may break into free-running oscillation under some unfavorable transient conditions, such as the turning on of the power supplies or the application of sudden heavy loads. Second, the measured values of filter gain, Q, and phase shift do not agree with the values calculated from these equations. These contradictory results can be explained if the three non-ideal operational amplifiers are taken into consideration. Real operational amplifiers normally have a single-pole gain and phase frequency response, and the transfer function is

$$T(s) = \frac{A_o \omega_c}{s + \omega_c}$$

For a typical low-cost operational amplifier, A_o is of the order of 100 dB and ω_c is 10 Hz. If $T(s)$ is included in the derivation of the filter's overall transfer function, the mathematics becomes manageable only by a computer. Figure 6.10 shows that if real operational amplifiers' gain and phase are taken into account, the filter Q increases with frequency rather fast beyond a certain point. Note that there is a maximum limit to the Q that is achievable even if $K_4 R_4$ (Fig. 6.8) is infinite.

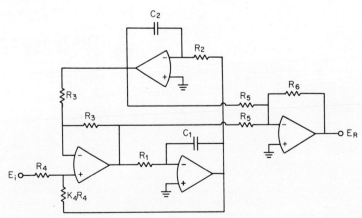

Fig. 6.9 Band-reject filter using the state-variable technique.

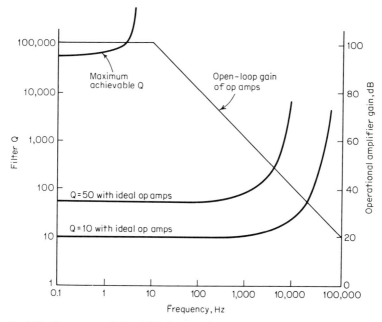

Fig. 6.10 Computer analysis of Q behavior, taking into account real operational amplifiers' gain and phase.

6.1.6 Universal linear-voltage-tunable active filter

The recent availability of low-cost IC multipliers has made voltage-tunable active filters economically practical. Adding two multipliers to the universal active filter creates a tunable filter, whose frequency varies linearly with the control voltage and which provides simultaneous low-pass, high-pass, and bandpass outputs. The circuit diagram is given in Fig. 6.11. The filter's Q factor and

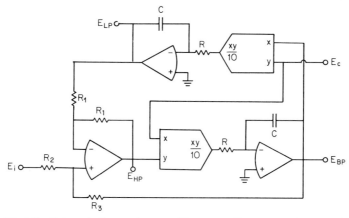

Fig. 6.11 Universal linear-voltage-tunable active filter, $\omega_o = E_c/10RC$ and $E_c > 0$ V.

passband gain remain constant while the frequency is varied. The voltage transfer functions between the respective outputs and the input are as follows:

$$\text{Low-pass} \quad \frac{E_{LP}(s)}{E_i(s)} = \frac{K_{LP}\omega_0^2}{s^2 + s\omega_0/Q + \omega_0^2}$$

$$\text{High-pass} \quad \frac{E_{HP}(s)}{E_i(s)} = \frac{K_{HP}s^2}{s^2 + s\omega_0/Q + \omega_0^2}$$

$$\text{Bandpass} \quad \frac{E_{BP}(s)}{E_i(s)} = \frac{K_{BP}s\omega_0/Q}{s^2 + s\omega_0/Q + \omega_0^2}$$

where $\omega_0 = \dfrac{E_c}{10RC}$

$$Q = \frac{R_2 + R_3}{2R_2}$$

$$K_{LP} = K_{HP} = \frac{2R_3}{R_2 + R_3}$$

$$K_{BP} = \frac{R_3}{R_2}$$

It should be noticed that the control voltage E_c must be positive; otherwise the complex conjugate poles will move to the right half plane. This means that only two-quadrant multipliers are needed. Voltage-tunable filters are widely used in such applications as spectrum analyzers, tracking and adaptive filters, vibration analyzers, voltage-controlled oscillators, and automatic test equipment.

6.1.7 All-pass filters An all-pass filter has all its zeros in the right half s plane, and all its poles are the left half plane images of its zeros. The transfer function of an all-pass filter may consist of simple and/or complex conjugate poles and zeros and is of the form

$$\frac{E_o(s)}{E_i(s)} = \frac{(s - \omega_1)(s^2 - s\omega_2/Q_2 + \omega_2)\ldots}{(s + \omega_1)(s^2 + s\omega_2/Q_2 + \omega_2^2)\ldots}$$

All-pass filters have constant unity magnitude gain for all frequency components of the input signal but provide predictable phase shifts for different frequencies of the input signal. These circuits are commonly used for the design of delay lines and phase correctors.

Figure 6.12 shows a first-order all-pass filter using one operational amplifier and a few passive components. The voltage transfer function is

$$\frac{E_o(s)}{E_i(s)} = -\frac{s - 1/R_1C_1}{s + 1/R_1C_1}$$

As the signal frequency varies from direct current to infinity, the phase shift will vary from 0 to $+180°$. Alternatively, if the signal frequency is kept constant, the filter's phase shift may be varied in the range between 0 to $+180°$ by adjusting either R_1 or C_1 or both. If the positions of R_1 and C_1 are interchanged, the new transfer function becomes

$$\frac{E_o(s)}{E_i(s)} = \frac{s - 1/R_1C_1}{s + 1/R_1C_1}$$

The phase-shift range is then from $-180°$ to $-0°$.

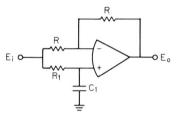

Fig. 6.12 A first-order all-pass filter that shifts from $0°$ to $+180°$. If the positions of R_1 and C_1 are interchanged, the all-pass filter will shift from $-180°$ to $-0°$.

A second-order all-pass filter with complex conjugate poles and zeros may be synthesized with a bandpass filter and a difference amplifier, as illustrated in Fig. 6.13. The transfer function of the bandpass filter is

$$\frac{V_o(s)}{E_i(s)} = \frac{s\omega_0/Q}{s^2 + s\omega_0/Q + \omega_0^2} \qquad (6\text{-}17)$$

The output amplifier takes the difference between E_i and V_o as follows:

$$E_o = E_i - 2V_o \qquad (6\text{-}18)$$

Eliminating V_o from Eqs. (6–17) and (6–18) results in the following all-pass transfer function:

$$\frac{E_o(s)}{E_i(s)} = \frac{s^2 - s\omega_0/Q + \omega_0^2}{s^2 + s\omega_0/Q + \omega_0^2}$$

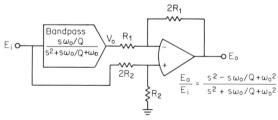

Fig. 6.13 Second-order all-pass filter with complex conjugate poles and zeros.

6.1.8 Third-order active filters with one operational amplifier

A third-order low-pass or high-pass filter can be designed with a two-pole section plus a one-pole RC section isolated by an operational amplifier. This approach requires two operational amplifiers to synthesize three poles. The mathematics are simple because the two sections are independent of each other. The same third-order filter can also be synthesized with one operational amplifier, although the calculations of component values may be somewhat time-consuming. If the quantity of filters required is high, the cost saving of one operational amplifier per filter produced may be well worth the effort and time spent on the computations.

Three-pole low-pass filter. The third-order low-pass active filter requiring only one operational amplifier is given in Fig. 6.14. The voltage transfer function in terms of circuit elements is

$$\frac{E_o(s)}{E_i(s)} = \frac{m}{\begin{array}{l} s^3 R_1 R_2 R_3 C_1 C_2 C_3 + s^2[R_2 R_3 C_2 C_3(1 + R_1/R_2) \\ \quad + R_1 R_3 C_1 C_3(1 + R_2/R_3) - R_1 R_2 C_1 C_2(m-1)] \\ \quad + s[R_1 C_1 + R_3 C_3(1 + R_2/R_3 + R_1/R_3) \\ \quad - R_2 C_2(1 + R_1/R_2)(m-1)] + 1 \end{array}} \quad (6\text{--}19)$$

The standard three-pole transfer function in terms of filter parameters is

$$\frac{E_o(s)}{E_i(s)} = \frac{K\omega_1\omega_2^2}{(s + \omega_1)(s^2 + s\omega_2/Q + \omega_2^2)}$$

$$= \frac{K}{s^3/\omega_1\omega_2^2 + s^2(1/Q\omega_1\omega_2 + 1/\omega_2^2) + s(1/\omega_1 + 1/Q\omega_2) + 1} \quad (6\text{--}20)$$

If the corresponding coefficients in Eqs. (6–19) and (6–20) are equated, the following four nonlinear equations result:

$$m = K \quad (6\text{--}21)$$

$$R_1 R_2 R_3 C_1 C_2 C_3 = \frac{1}{\omega_1 \omega_2^2} \quad (6\text{--}22)$$

$$R_2 R_3 C_2 C_3\left(1 + \frac{R_1}{R_2}\right) + R_1 R_3 C_1 C_3\left(1 + \frac{R_2}{R_3}\right) - R_1 R_2 C_1 C_2(m-1)$$
$$= \frac{1}{Q\omega_1\omega_2} + \frac{1}{\omega_2^2} \quad (6\text{--}23)$$

$$R_1 C_1 + R_3 C_3\left(1 + \frac{R_2}{R_3} + \frac{R_1}{R_3}\right) - R_2 C_2\left(1 + \frac{R_1}{R_2}\right)(m-1) = \frac{1}{\omega_1} + \frac{1}{Q\omega_2} \quad (6\text{--}24)$$

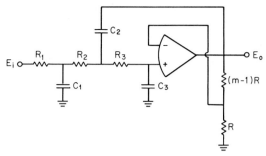

Fig. 6.14 Synthesize a three-pole low-pass filter with one amplifier.

To design a filter with given ω_1, ω_2, and Q, simply make $C_1 = C_2 = C_3 = C$ and $R_1 = R_2$. Equations (6-22), (6-23), and (6-24) then become

$$R_1^2 R_3 = \frac{1}{C^3 \omega_1 \omega_2^2}$$

$$R_1^2(2-m) + 3R_1 R_3 = \frac{1}{C^2}\left[\frac{1}{Q\omega_1\omega_2} + \frac{1}{\omega_2^2}\right]$$

$$R_1(5-2m) + R_3 = \frac{1}{C}\left[\frac{1}{\omega_1} + \frac{1}{Q\omega_2}\right]$$

The above three equations can be solved for the three unknowns, R_1, R_3, and m, in terms of ω_1, ω_2, Q, and C.

Design procedures. ω_1, ω_2, and Q are given parameters. Choose a suitable value for $C(=C_1=C_2=C_3)$. To calculate the values for $R_1(=R_2)$, R_3, and m, do the following computations step by step with a desk calculator. The computations can also be programmed into a computer for repeated usage.

1. $A_1 = \dfrac{1}{C^3 \omega_1 \omega_2^2}$

2. $A_2 = \dfrac{1}{C^2}\left(\dfrac{1}{Q\omega_1\omega_2} + \dfrac{1}{\omega_2^2}\right)$

3. $A_3 = \dfrac{1}{C}\left(\dfrac{1}{\omega_1} + \dfrac{1}{Q\omega_2}\right)$

4. $P = -A_3$
5. $Q = 2A_2$
6. $R = -5A_1$
7. $B_1 = \dfrac{1}{3}(3Q - P^2)$

8. $B_2 = \dfrac{1}{27}(2P^3 - 9PQ + 27R)$

9. $G_1 = \left(\dfrac{B_2{}^2}{4} + \dfrac{B_1{}^3}{27}\right)^{1/2}$

10. $H_1 = \left(-\dfrac{B_2}{2} + G_1\right)^{1/3}$

11. G_2 = absolute value of $\left(-\dfrac{B_2}{2} - G_1\right)$

12. $H_2 = -G_2{}^{1/3}$

13. $R_1 = H_1 + H_2 - \dfrac{P}{3}$

14. $S = \dfrac{A_1}{R_1{}^3}$

15. $R_3 = SR_1$

16. $m = 2 + 3S - \dfrac{A_2}{R_1{}^2}$

Three-pole high-pass filter. Figure 6.15 illustrates the circuit connections for a third-order high-pass filter using one operational amplifier. The voltage transfer function is

$$\dfrac{E_o(s)}{E_i(s)} = \dfrac{ms^3 R_1 R_2 R_3 C_1 C_2 C_3}{\begin{array}{l} s^3 R_1 R_2 R_3 C_1 C_2 C_3 + s^2 [R_1 R_2 C_2 C_3 + R_2 R_3 C_2 C_3] \\ + R_1 R_2 C_1 (C_2 + C_3) - (m-1) R_1 R_3 C_3 (C_1 + C_2)] \\ + s[R_1 C_1 + R_1 C_2 + R_2 C_2 + R_2 C_3 - (m-1) R_3 C_3] + 1 \end{array}}$$

(6–25)

The transfer function of a third-order high-pass filter in terms of filter parameters is

$$\dfrac{E_o(s)}{E_i(s)} = \dfrac{Ks^3}{(s + \omega_1)(s^2 + s\omega_2/Q + \omega_2{}^2)}$$

$$= \dfrac{Ks^3/(\omega_1 \omega_2{}^2)}{s^3/\omega_1 \omega_2{}^2 + s^2(1/Q\omega_1 \omega_2 + 1/\omega_2{}^2) + s(1/\omega_1 + 1/Q\omega_2) + 1}$$

(6–26)

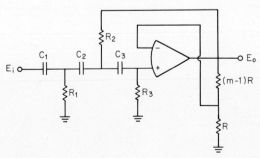

Fig. 6.15 Synthesize a three-pole high-pass filter with one amplifier.

If the corresponding coefficients in Eqs. (6–25) and (6–26) are equated, a set of four simultaneous nonlinear equations results. Making $C_1 = C_2 = C_3 = C$ and $R_1 = R_2$ means that R_1, R_3, and m can be solved for in terms of ω_1, ω_2, Q, and C in the same way as in the low-pass case.

Design procedures. ω_1, ω_2, and Q are given high-pass filter parameters. Choose a suitable value for $C(=C_1=C_2=C_3)$. To calculate the values for $R_1(=R_2)$, R_3, and m, do the following step-by-step computations:

1. $A_1 = \dfrac{1}{C^3 \omega_1 \omega_2^2}$

2. $A_2 = \dfrac{1}{C^2}\left(\dfrac{1}{Q\omega_1\omega_2} + \dfrac{1}{\omega_2^2}\right)$

3. $A_3 = \dfrac{1}{C}\left(\dfrac{1}{\omega_1} + \dfrac{1}{Q\omega_2}\right)$

4. $P = -\dfrac{2A_3}{5}$

5. $Q = \dfrac{A_2}{5}$

6. $R = -\dfrac{A_1}{5}$

7. $B_1 = \dfrac{1}{3}(3Q - P^2)$

8. $B_2 = \dfrac{1}{27}(2P^3 - 9PQ + 27R)$

9. $G_1 = \left(\dfrac{B_2^2}{4} + \dfrac{B_1^3}{27}\right)^{1/2}$

10. $H_1 = \left(-\dfrac{B_2}{2} + G_1\right)^{1/3}$

11. $G_2 =$ absolute value of $\left(-\dfrac{B_2}{2} - G_1\right)$

12. $H_2 = -G_2^{1/3}$

13. $R_1 = H_1 + H_2 - \dfrac{P}{3}$

14. $R_3 = \dfrac{A_1}{R_1^2}$

15. $m = \dfrac{4R_1 + R_3 - A_3}{R_3}$

6.1.9 Computer-aided design of fourth-order low-pass filter with one operational amplifier

The network configuration of a fourth-order low-pass active RC filter is shown in Fig. 6.16. The filter transfer function is

$$\frac{E_o(s)}{E_i(s)} = \frac{m}{s^4\omega_4 + s^3\omega_3 + s^2\omega_2 + s\omega_1 + 1} \quad (6\text{-}27)$$

where $\omega_4 = R_1 R_2 R_3 R_4 C_1 C_2 C_3 C_4$

$$\omega_3 = C_2 C_3 C_4 R_3 R_4 (R_1 + R_2) + C_1 R_1 \left[C_3 C_4 R_3 R_4 \left(1 + \frac{R_2}{R_3}\right) + C_2 R_2 Z \right]$$

$$\omega_2 = \left[C_3 C_4 R_3 R_4 \left(1 + \frac{R_2}{R_3}\right) + C_2 R_2 Z \right]\left(1 + \frac{R_1}{R_2}\right) + C_1 R_1 \left[Z\left(1 + \frac{R_2}{R_3}\right) \right.$$
$$\left. + C_2 R_2 - \frac{C_4 R_4 R_2}{R_3} \right] - \frac{C_3 C_4 R_3 R_4 R_1}{R_2}$$

$$\omega_1 = \left[Z\left(1 + \frac{R_2}{R_3}\right) + C_2 R_2 - \frac{C_4 R_4 R_2}{R_3} \right]\left(1 + \frac{R_1}{R_2}\right) + C_1 R_1 (1 - m) - Z\frac{R_1}{R_2}$$

$$Z = (1 - m)C_3 R_3 + C_4 R_4 + C_4 R_3$$

The transfer function of the fourth-order low-pass filter in terms of two quadratic filter parameters is

$$\frac{E_o(s)}{E_i(s)} = \frac{K\omega_1^2 \omega_2^2}{(s^2 + s\omega_1/Q_1 + \omega_1^2)(s^2 + s\omega_2/Q_2 + \omega_2^2)}$$

$$= \frac{K}{s^4/(\omega_1 \omega_2)^2 + s^3(1/Q_1\omega_1\omega_2^2 + 1/Q_2\omega_1^2\omega_2)}$$
$$+ s^2(1/\omega_1^2 + 1/(Q_1 Q_2 \omega_1 \omega_2) + 1/\omega_2^2)$$
$$+ s(1/Q_1\omega_1 + 1/Q_2\omega_2) + 1 \quad (6\text{-}28)$$

If we equate the corresponding coefficients in Eqs. (6-27) and (6-28), we have $m = K$ and a set of four simultaneous nonlinear equations in terms of unknown variables ($R_1, R_2, R_3, R_4, C_1, C_2, C_3, C_4$, and m) and known filter

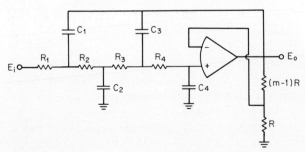

Fig. 6.16 Fourth-order low-pass active filter with one operational amplifier.

parameters (ω_1, Q_1, ω_2, and Q_2). Since there are nine unknowns but only four simultaneous equations, the solution will not be unique. One practical design is to make $R_2 = R_3$ and $C_1 = C_2 = C_3 = C_4 = C$. Choose a suitable value for C and solve for R_1, $R_2 (= R_3)$, R_4, and m in terms of C, ω_1, Q_1, ω_2 and Q_2 from the set of four equations, simplified as follows:

$$\frac{1}{R_1} + \frac{4}{R_2} + \frac{2-m}{R_4} = A_1 \qquad (6\text{--}29)$$

$$\frac{3}{R_1 R_2} + \frac{2-m}{R_1 R_4} + \frac{3}{R_2^2} + \frac{7-3m}{R_2 R_4} = A_2 \qquad (6\text{--}30)$$

$$\frac{1}{R_1 R_2^2} + \frac{5-2m}{R_1 R_2 R_4} + \frac{2(2-m)}{R_2^2 R_4} = A_3 \qquad (6\text{--}31)$$

$$\frac{1}{R_1 R_2^2 R_4} = A_4 \qquad (6\text{--}32)$$

where $A_1 = C \left(\dfrac{\omega_1}{Q_1} + \dfrac{\omega_2}{Q_2} \right)$

$A_2 = C^2 \left(\dfrac{\omega_1 \omega_2}{Q_1 Q_2} + \omega_1^2 + \omega_2^2 \right)$

$A_3 = C^3 \left(\dfrac{\omega_1 \omega_2}{Q_1} + \dfrac{\omega_1^2 \omega_2}{Q_2} \right)$

$A_4 = C^4 \omega_1^2 \omega_2^2$

If R_4 and m are eliminated from Eqs. (6–29), (6–30), (6–31), and (6–32), the following two simultaneous nonlinear equations in terms of unknown variable $P_1 (= 1/R_1)$ and $P_2 (= 1/R_2)$ result:

$$P_1^2 + P_1 \frac{9P_2 - 2A_1}{2} + \frac{8P_2^4 - 2A_1 P_2^3 + A_3 P_2 - A_4}{2P_2^2} = 0$$

$$P_1 = \frac{-8P_2^5 + 2A_1 P_2^4 - A_3 P_2^2 + 5A_4 P_2}{29 P_2^4 - 10 A_1 P_2^3 + 4 A_2 P_2^2 - 2 A_3 P_2 + 2 A_4}$$

If we try to go further solving for P_1 and P_2 analytically, the mathematics becomes too tedious to handle. However, if a programmable desk calculator or a computer is available, the two simultaneous nonlinear equations can be solved by numerical methods.

The computer program WONGA to design the fourth-order filter is listed in Fig. 6.17a. The program will ask for input data "AF1, F1, AF2, F2, C" and "R2ST, R2SP," where AF1 = $1/Q$, F1 (in hertz) = $\omega_1/2\pi$, AF2 = $1/Q_2$, F2 (in hertz) = $\omega_2/2\pi$, C (in microfarads) = $C_1 = C_2 = C_3 = C_4$, and R2ST

```
100-    WONGA    4 POLE 1 AMPLIFIER DESIGN    R IN KOHMS
110     DIMENSION P1(3),P2(3),ANP(3),DP(3),PP(3)
120     10 PRINT,↑↑,"AF1,F1,AF2,F2,C",
130     INPUT,AF1,F1,AF2,F2,C
140     C=C*1.E-3
150     PRINT,↑,"R2ST,R2SP",
160     INPUT,R2ST,R2SP
170     P2ST=1./R2SP
180     P2SP=1./R2ST
190     F1=2.*3.14159*F1
200     F2=2.*3.14159*F2
210     AK1=(AF1*F1+AF2*F2)*C
220     AK2=(AF1*AF2*F1*F2+F1*F1+F2*F2)*(C*C)
230     AK3=(AF1*F1*F2*F2+AF2*F1*F1*F2)*C*C*C
240     AK4=F1*F1*F2*F2*C*C*C*C
250     P2(1)=P2ST
260     P2(2)=(P2ST+P2SP)/2.
270     P2(3)=P2SP
280     20 DO 100 K1=1.3
290     ANP(K1)=-8.*(P2(K1)**5)+2.*AK1*(P2(K1)**4)
300     ANP(K1)=ANP(K1)-AK3*P2(K1)*P2(K1)+5.*AK4*P2(K1)
310     DP(K1)=29.*(P2(K1)**4)-10.*AK1*(P2(K1)**3)+4.*AK2*(P2(K1)**2)
320     DP(K1)=DP(K1)-2.*AK3*P2(K1)+2.*AK4
330     P1(K1)=ANP(K1)/DP(K1)
340     PP(K1)=P1(K1)*P1(K1)+.5*P1(K1)*(9.*P2(K1)-2.*AK1)
350     PP(K1)=PP(K1)+(8.*(P2(K1)**4)-2.*AK1*(P2(K1)**3)
360   + +AK3*P2(K1)-AK4)/(2.*P2(K1)*P2(K1))
370     100 CONTINUE
380     PPA=PP(1)*PP(2)
390     IF(PPA) 120,200,110
400     110 PPA=PP(2)*PP(3)
410     IF(PPA) 160,200,150
420     150 GOTO 10
430     120 P2(3)=P2(2)
440     P2(2)=(P2(1)+P2(2))/2.
450     X=.000001
460     P=ABS((P2(2)-P2(3))/P2(2))-X
470     IF(P) 200,200,20
480     160 P2(1)=P2(2)
490     P2(2)=(P2(2)+P2(3))/2.
500     P=ABS((P2(2)-P2(3))/P2(2))-X
510     IF(P) 200,200,20
520     200 R2=1./P2(2)
530     R3=R2
540     R1=1./P1(2)
550     R4=1./(R1*R2*R3*AK4)
560     AK=(1./R1+4./R2-AK1)*R4+2.
570     PRINT 210,R1,R2,R3,R4,AK
580     210 FORMAT(/,"R1= ",E11.5,6X,"R2= ",E11.5,6X,"R3= ",
590   + E11.5,/,"R4= ",E11.5,6X,"K= ",E11.5)
600     GOTO 10
```

Fig. 6.17a Program WONGA to design fourth-order low-pass active filter with one operational amplifier. The program will ask for AF1, F1, AF2, F2, C, and R2ST, R2SP, and will print out values of R_1, R_2, R_3, R_4, and K.

```
1000    C CIRCUIT
1010    DIMENSION A(100,5),B(50),YA(60,60),YB(60,60)
1020    COMMON W(120,120),BB(120)
1030    DOUBLE PRECISION W,BB
1040    10 PRINT 10000
1050    10000 FORMAT (1X, "NO. OF NODES, NO. OF COMPONENTS")
1060    READ/,M1,M
1070    PRINT 10001
1080    10001 FORMAT (1X, "CURRENT SOURCE INPUT")
1090    READ/,BBB
1100    PRINT 10002
1110    10002 FORMAT (1X, "A(1,1),A(1,2),. . ., A(1,5)")
1120    PRINT 10003
1130    10003 FORMAT (1X, " . . . . . . . .")
1140    PRINT 10004
1150    10004 FORMAT (1X, "A(M,1),A(M,2), . . ., A(M,5)")
1160    READ/,((A(I1,I2),I2=1,5),I1=1,M)
1170    20 PRINT 10005
1180    10005 FORMAT (1X, "NO. OF FREQ, F(1),. ., F(M2)")
1190    READ/,M2,(B(K5),K5=1,M2)
1200    PRINT 15
1210    15 FORMAT (1X,//,3X,"FREQ(HZ)",6X,"GAIN(DB)",6X,"PHASE(DEG)",/)
1220    MN=1
1230    DO 100 K5=1,M2
1240    B(K5)=2.*3.14159*B(K5)
1250    55 DO 110 K1=1,M1
1260    DO 120 K2=1,M1
1270    YA(K1,K2)=0.
1280    YB (K1,K2)=0.
1290    120 CONTINUE
1300    110 CONTINUE
1310    DO 150 I1=1,M
1320    NN=A(I1,3)
1330    GOTO(160,170,180,190),NN
1340    160 N1=A(I1,1)
1350    N2=A(I1,2)
1360    IF(N1) 300,310,300
1370    300 IF(N2) 320,330,320
1380    310 YA(N2,N2)=YA(N2,N2)+1./A(I1,4)
1390    GOTO 150
1400    330 YA(N1,N1)=YA(N1,N1)+1./A(I1,4)
1410    GOTO 150
1420    320 YA(N1,N1)=YA(N1,N1)+1./A(I1,4)
1430    YA(N2,N2)=YA(N2,N2)+1./A(I1,4)
1440    YA(N1,N2)=YA(N1,N2)-1./A(I1,4)
1450    YA(N2,N1)=YA(N2,N1)-1./A(I1,4)
1460    GOTO 150
1470    170 N1=A(I1,1)
1480    N2=A(I1,2)
1490    IF(N1) 340,350,340
1500    340 IF(N2) 360,370,360
```

236 Function Circuits

```
AF1,F1,AF2,F2,C? 1.8478,1.,.7654,1.,1.

R2ST,R2SP? 100.,800.

R1=   .11942E+03        R2=   .42820E+03        R3=   .42820E+03
R4=   .29303E+02        K=    .20380E+01


AF1,F1,AF2,F2,C? 1.915954,.962319,1.2414,1.0789,1.

R2ST,R2SP? 100.,800.

R1=   .10695E+03        R2=   .42747E+03        R3=   .42747E+03
R4=   .30457E+02        K=    .19606E+01


AF1,F1,AF2,F2,C? 1.418218,.597002,.340072,1.03127,1.

R2ST,R2SP? 100.,800.

R1=   .22457E+03        R2=   .56007E+03        R3=   .56007E+03
R4=   .24030E+02        K=    .20978E+01


AF1,F1,AF2,F2,C? 1.075906,.470711,.217681,.963678,1.

R2ST,R2SP? 100.,800.

R1=   .33962E+03        R2=   .60972E+03        R3=   .60972E+03
R4=   .24697E+02        K=    .21236E+01


AF1,F1,AF2,F2,C? STOP
```

Fig. 6.17b Data printout of four designs: the Butterworth, the Bessel, the 0.5 dB Tchebysheff, and the 2 dB Tchebysheff.

and R2SP are the estimated start and stop values of R_2 in kilohms, in between which a solution for R_2 exists. R2ST and R2SP can be set as far apart as you want at the expense of more computer time. The program will print out values for R_1, R_2, R_3, and R_4, all in kilohms, and the value for $K\ (=m)$. Figure 6.17b shows the output data printout of the program for four different fourth-order low-pass designs: the Butterworth, the Bessel, the 0.5 dB Tchebysheff, and the 2 dB Tchebysheff. In the first portion of the data, where the Butterworth design with -3 dB frequency at 1 Hz is listed, AF1 = 1.8478, F1 = 1 Hz, AF2 = 0.7654, F2 = 1 Hz, and C is chosen to be 1 μF. R2ST and R2SP are loosely estimated at 100 and 800 kΩ, respectively. The calculated resistor values are printed out as R1 = 119.42 kΩ, R2 = 428.20 kΩ, R3 = 428.20 kΩ, and R4 = 29.303 kΩ, and K is 2.038. The rest of the data printout can be interpreted similarly. The Bessel design has a phase shift of π radians at 1 Hz, while the two

Tchebysheff designs are such that the gain curve first departs from their respective maximum ripple band at 1 Hz.

6.2 Butterworth, Bessel, and Tchebysheff Characteristics

In this section, we will discuss the three most popular approximation techniques: Butterworth, Bessel, and Tchebysheff. Although the discussions will concentrate on low-pass filters, the basic characteristics (e.g., equal ripple for Tchebysheff approximations) apply as well to high-pass, bandpass, and band-reject filters. An ideal low-pass filter (sometimes called a brick-wall filter) is a network which has a constant amplitude characteristic from direct current up to its cutoff frequency ω_o and has zero output for any frequencies above ω_o. Certainly, such a filter is not realizable, and, therefore, an approximation of the ideal filter response is required.

Low-pass Butterworth characteristics. Butterworth (sometimes called maximally flat) filters have excellent gain accuracy in the lower portion of the passband and reasonably well-behaved phase-shift characteristics. They provide the flattest possible amplitude response obtainable without having gain ripple in the pass- and stop bands. The attenuation rate beyond the passband is set by the number of poles: for N poles the rolloff is $6N$ dB per octave or $20N$ dB per decade. The phase characteristics are not too linear. The output overshoots in response to a step input, and the overshoot increases with the number of poles. Amplitude and phase response curves of Butterworth filters are given in Fig. 6.18.

The poles of Butterworth filters lie in a unit circle in the complex s plane. Expressed in single-order and second-order quadratic form, the individual filter section's center frequency and Q are tabulated in Table 6.1 for up to $N = 8$. The parameters are normalized to be -3 dB down at $\omega = 1$ rad/s. The magnitude transfer function of Butterworth filters can be simplified as

$$\left|\frac{E_o}{E_i}(j\omega)\right| = \frac{K}{\sqrt{1+\omega^{2N}}}$$

where K is the gain at direct current and N is the number of poles. Obviously, at $\omega = 1$, $|E_o/E_i| = K/\sqrt{2}$, i.e., -3 dB down.

Low-pass Bessel characteristics. The Bessel filter provides excellent phase-shift linearity, but the amplitude cutoff is not as sharp as with Butterworth or Tchebysheff filters. Bessel filters, as a group, are also referred to as a type of "linear phase" filter. These filters will pass rectangular pulses with a minimum of distortion and with a delay time that is linearly

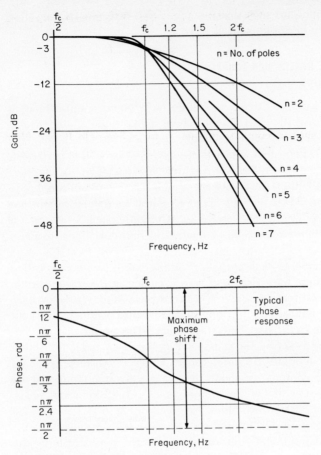

Fig. 6.18 Low-pass Butterworth response.

proportional to the phase-shift characteristics. That is, the time delay through the filter is almost constant with frequency and is equal to the slope of the filter phase characteristic. The overshoot to a step input is essentially zero for Bessel filters, where Tchebysheff filters may exhibit more than 25 percent overshoot. Because of these characteristics, Bessel filters are sometimes used to provide time delays. Bessel filters are also used for low-pass filtering of rectangular waveforms in pulsewidth modulation systems, voltage-to-frequency converters, and similar circuits. "Running average" filters often require a Bessel-type response.

Since phase shift is the parameter of interest for a Bessel filter, the cutoff frequency f_c is defined in terms of phase shift. The frequency at which the phase shift is one-half the maximum phase shift is defined as the cutoff frequency f_c. The maximum phase shift depends directly upon the order (number of poles) of the filter. For a Bessel filter of N poles and phase shift $\theta(f)$,

$$\theta(f_c) = \frac{\theta(f)_{max}}{2} = \frac{N\pi/2}{2} \quad \text{rad}$$

$$= N\frac{\pi}{4} \quad \text{rad}$$

Thus the phase shift at cutoff frequency f_c for a five-pole Bessel filter would be $5\pi/4$ radians. The phase shift at cutoff frequency, $\theta(f_c)$, is often referred to as the "phase constant," or "delay at cutoff." For accurate delay (phase constant), f_c should be about twice as high as the maximum signal frequency. Figure 6.19 shows some representative amplitude and phase response curves for Bessel filters. Note that the curves are plotted

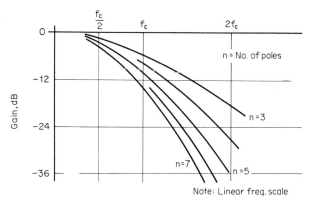

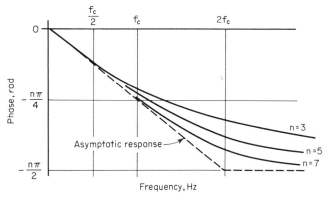

Fig. 6.19 Bessel (linear phase) response curves.

TABLE 6.1 Low-pass Butterworth Filter Parameters, −3 dB at $\omega_c = 1$ rad/s

N	ω_1	Q_1	ω_2	Q_2	ω_3	Q_3	ω_4	Q_4
2	1.00000	0.707107						
3	1.00000	1.00000	1.00000	Real pole				
4	1.00000	1.30656	1.00000	0.541196				
5	1.00000	1.61803	1.00000	0.618034	1.00000	Real pole		
6	1.00000	1.93185	1.00000	0.707107	1.00000	0.517638		
7	1.00000	2.24698	1.00000	0.801938	1.00000	0.554958	1.00000	Real pole
8	1.00000	2.56291	1.00000	0.899977	1.00000	0.601345	1.00000	0.509796

on a linear rather than a logarithmic frequency scale. Filter parameters up to $N = 8$ are given in Table 6.2.

Low-pass Tchebysheff characteristics. Filters of this class offer a maximum attenuation rate beyond cutoff, which is achieved at the expense of ripples in the passband. Higher attenuation rates outside the passband are accompanied by correspondingly greater passband ripple. The step response of such filters is somewhat damped in comparison to Butterworth or Bessel filters and will exhibit considerable overshoot and ringing. These types of filters are sometimes called *equal-ripple filters*. Tchebysheff filters are excellent for many audio applications and others where ripple in the passband is not important, but where sharp cutoff is required. In general, Tchebysheff filters should not be used where good transient response is important.

Cutoff frequency f_c is the frequency at which the gain curve first departs from the maximum ripple band. Response curves for 0.5 dB and 2 dB ripple Tchebysheff filters are given in Fig. 6.20. Note that the slope just outside the cutoff frequency exceeds $20N$ dB per decade. It will, however, reach $20N$ dB per decade asymptotes at high frequencies. Filter parameters are tabulated in Tables 6.3 and 6.4 for 0.5 dB and 2 dB ripple filters, respectively.

TABLE 6.2 Low-pass Bessel Filter Parameters; the Phase Shift at $\omega_c = 1$ rad/s Is $N\pi/4$ rad

N	ω_1	Q_1	ω_2	Q_2	ω_3	Q_3	ω_4	Q_4
2	1.00000	0.577350						
3	1.07869	0.691047	0.985560	Real pole				
4	1.07890	0.805538	0.962319	0.521935				
5	1.08504	0.916478	0.962003	0.563536	0.928640	Real pole		
6	1.09270	1.02331	0.969010	0.611195	0.920141	0.510318		
7	1.10034	1.12626	0.978443	0.660821	0.921478	0.532356	0.904336	Real pole
8	1.10046	1.22567	0.982040	0.710853	0.921150	0.559609	0.894187	0.505991

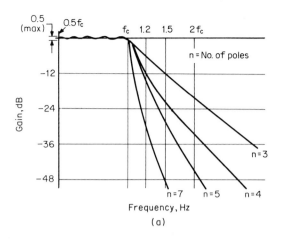

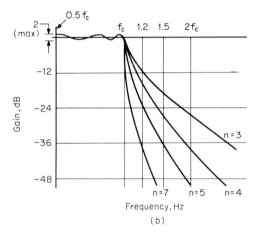

Fig. 6.20 Tchebysheff response curves: (a) 0.5 dB passband ripple; and (b) 2 dB passband ripple.

TABLE 6.3 Low-pass 0.5 dB Ripple Tchebysheff Filter Parameters; $\omega_c = 1$ rad/s Is the Frequency at Which the Gain Curve First Departs from the 0.5 dB Ripple Band

N	ω_1	Q_1	ω_2	Q_2	ω_3	Q_3	ω_4	Q_4
2	1.23134	0.863721						
3	1.06885	1.70619	0.626456	Real pole				
4	1.03127	2.94055	0.597002	0.705110				
5	1.01774	4.54496	0.690483	1.17781	0.362320	Real pole		
6	1.01145	6.51283	0.768121	1.81038	0.396229	0.683639		
7	1.00802	8.84181	0.822729	2.57555	0.503863	1.09155	0.256170	Real pole
8	1.00595	11.5308	0.861007	3.46568	0.598874	1.61068	0.296736	0.676575

TABLE 6.4 Low-pass 2 dB Ripple Tchebysheff Filter Parameters; $\omega_c = 1$ rad/s Is the Frequency at Which the Gain Curve First Departs from the 2 dB Ripple Band

N	ω_1	Q_1	ω_2	Q_2	ω_3	Q_3	ω_4	Q_4
2	0.907227	1.12865						
3	0.941326	2.55164	0.368911	Real pole				
4	0.963678	4.59388	0.470711	0.929449				
5	0.975790	7.23228	0.627017	1.77509	0.218308	Real pole		
6	0.982828	10.4616	0.730027	2.84426	0.316111	0.901595		
7	0.987226	14.2802	0.797114	4.11507	0.460853	1.64642	0.155340	Real pole
8	0.990141	18.6873	0.842486	5.58354	0.571925	2.532267	0.237699	0.892354

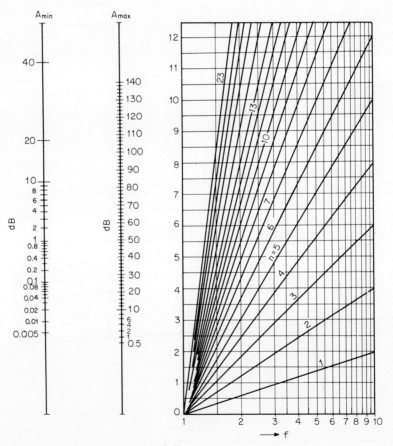

Fig. 6.21 Nomograph for Butterworth filters.

6.3 Estimating Filter Complexity

Nomographs for Butterworth and Tchebysheff filters are given in Figs. 6.21 and 6.22, respectively. Originally developed by Kawakami, they enable the user to estimate the number of sections required to approximate a given magnitude response curve. The nomographs may be used for low-pass, high-pass, and bandpass as well as band-reject filters. To estimate the filter complexity, first determine the acceptable passband ripple, A_{min}, and the maximum stop-band attenuation, A_{max}. Figure 6.23 presents a few examples. Then calculate f, the ratio of stop-band bandwidth (f_{max}) to passband ripple bandwidth (f_{min}) (for high-pass and band-reject filters, $f = f_{min}/f_{max}$). To use the nomograph, enter the passband ripple at the left straight line (X_1) and the stop-band attenuation at the right straight line (X_2). Draw a line between X_1 and X_2, and extend it to the left-hand side of the nomograph (X_3). Draw a vertical line at X_4 and a horizontal

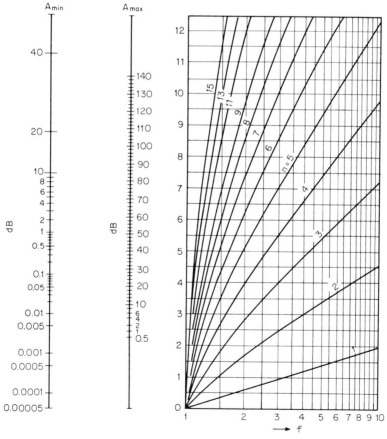

Fig. 6.22 Nomograph for Tchebysheff filters.

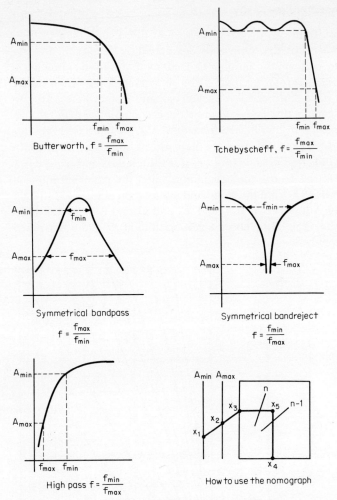

Fig. 6.23 Illustration examples to estimate filter complexity using Butterworth and Tchebysheff nomographs.

line at X_3 until they intersect at X_5. The next curve to the left of X_5 indicates the number of required poles for low-pass and high-pass filters or the number of required pole pairs for bandpass and band-reject filters. Note that the magnitude curve must be geometric-mean symmetrical about a center frequency for bandpass and band-reject filters (see Secs. 6.1.3 and 6.1.4).

6.4 Transformation of Transfer Functions

Every low-pass filter defines an equivalent high-pass, symmetrical bandpass, and symmetrical band-reject filter whose properties are related to those of the low-pass filter in a predictable manner. Therefore, to design high-pass, bandpass, or band-reject filters, it is advisable to first transform the given specifications to the low-pass equivalent. This is called the *low-pass prototype*. From readily available data, the transfer function of the low-pass prototype is then determined. In this section, we will discuss several transformation equations which change the low-pass prototype transfer function back to its equivalent high-pass, symmetrical bandpass, and symmetrical band-reject, respectively.

6.4.1 From low-pass to high-pass

Consider the transfer function of a given low-pass filter:

$$\frac{E_o(s)}{E_i(s)} = \frac{K}{(s + \omega_1)(s^2 + s\omega_2/Q_2 + \omega_2^2) \ldots} \quad (6\text{-}33)$$

The filter may consist of any number of simple poles and/or complex conjugate poles. The transfer function of an equivalent high-pass filter may be generated with the following transformation:

$$s = \frac{1}{p} \quad (6\text{-}34)$$

Substituting Eq. (6–34) into (6–33) and simplifying, the high-pass transfer function becomes

$$\frac{E_o(p)}{E_i(p)} = \frac{Kp^3/\omega_1\omega_2^2}{(p + 1/\omega_1)(p^2 + p/Q_2\omega_2 + 1/\omega_2^2) \ldots} \quad (6\text{-}35)$$

As shown in Fig. 6.24, suppose the original low-pass filter is X_2 dB down at its cutoff frequency of $\omega = 1$ rad/s, X_1 dB down at $\omega = \omega_1$, and X_3 dB down at $\omega = \omega_3$; then the transformed high-pass filter described by Eq. (6–35) will also be X_2 dB down at $\omega = 1$ rad/s but will be X_1 dB down at $\omega = 1/\omega_1$ and X_3 dB down at $\omega = 1/\omega_3$.

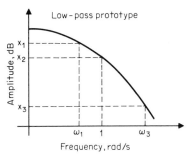
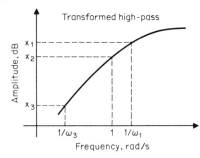

Fig. 6.24 Transformation from low-pass to high-pass using $s = 1/p$.

6.4.2 From low-pass to bandpass From a given low-pass transfer function, a geometric-mean symmetrical bandpass transfer function may be obtained by the following transformation equation:

$$s = p + \frac{1}{p} \qquad (6\text{--}36)$$

Substituting Eq. (6–36) into (6–33) and simplifying, the equivalent bandpass transfer function becomes

$$\frac{E_o(p)}{E_i(p)} = \frac{Kp^3}{(p^2 + p\omega_1 + 1)[p^4 + p^3\omega_2/Q_2 + p^2(2 + \omega_2^2) + p\omega_2/Q_2 + 1] \cdots} \qquad (6\text{--}37)$$

The quartic term in the denominator of Eq. (6–37) can be broken into two quadratic terms. Therefore, Eq. (6–37) may be rewritten as

$$\frac{E_o(p)}{E_i(p)} = \frac{Kp^3}{(p^2 + p\omega_a/Q_a + \omega_a^2)(p^2 + p\omega_b/Q_b + \omega_b^2)(p^2 + p\omega_c/Q_c + \omega_c^2) \cdots}$$

where $\omega_a = 1$

$$Q_a = \frac{1}{\omega_1}$$

$$Q_b = \sqrt{\frac{2}{4 + \omega_2^2 - A}} \qquad A = \sqrt{(4 + \omega_2^2)^2 - \frac{4\omega_2^2}{Q_2^2}} \qquad \text{for } Q_2 < 20$$

$$Q_b = \frac{4}{\omega_2} \sqrt{\frac{Q_2^2}{4 - (\omega_2 Q_2)^2}} \qquad \text{for } Q_2 \geq 20$$

$$\omega_b = \frac{1}{2}\left[\frac{Q_b \omega_2}{Q_b} + \sqrt{\left(\frac{Q_b \omega_2}{Q_2}\right)^2 - 4}\right]$$

$$Q_c = Q_b$$

$$\omega_c = \frac{1}{\omega_b}$$

It should be cautioned that in computing A, Q_b, and ω_b, a substantial number of significant digits should be carried on or false results will be obtained. For example, to transform $\omega_2 = 0.102921$ and $Q_2 = 2.94055$, the correct results are $A = 4.009981782$, $Q_b = 57.2152$, and $\omega_b = 1.05199$. However, if only six significant digits are retained for A (i.e., $A = 4.00998$), the calculated value for Q_b will be 57.1320, and this will make ω_b a complex number.

For every simple pole in the low-pass transfer function, a pair of complex conjugate poles are generated in the transformed bandpass transfer function, and for every pair of complex conjugate poles in the low-pass transfer function, two pairs of complex conjugate poles with the same Q factor are

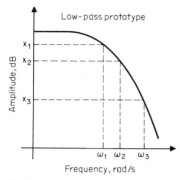 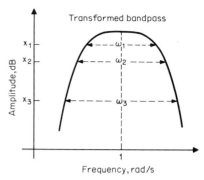

Fig. 6.25 Transformation from low-pass to bandpass using $s = p + 1/p$.

generated in the bandpass transfer function. The total number of poles is actually doubled in the transformation.

The amplitude relationship between the low-pass prototype and the transformed bandpass filter is illustrated in Fig. 6.25. An attenuation of X_n dB down at ω_n rad/s in the low-pass prototype becomes the X_n dB bandwidth of ω_n rad/s in the bandpass filter. The center frequency of the transformed bandpass filter is 1 rad/s.

6.4.3 From low-pass to band-reject

The transformation equation that changes a low-pass filter to a geometric-mean symmetrical band-reject filter is

$$\frac{1}{s} = p + \frac{1}{p} \qquad (6\text{--}38)$$

The term $p + 1/p$ on the right-hand side of Eq. (6–38) is the same as the right-hand side of the transformation equation from low-pass to bandpass, while the term $1/s$ on the left-hand side of Eq. (6–38) corresponds to the transformation equation from low-pass to high-pass. In other words, to derive the band-reject transfer function, first transform the low-pass prototype transfer function to high-pass, then use the formulas established in the preceding section to change the intermediate high-pass transfer function to band-reject.

6.5 A Practical Multistage Active Filter Design Example

Bandpass filters are considered by many to be one of the most tedious and time-consuming design jobs. A practical example is presented in this section to illustrate the step-by-step procedures for designing a high-performance eight-pole stagger-tuned bandpass filter. By following the same procedures with logical modifications, one can design high-pass and band-reject networks accordingly.

6.5.1 Given specifications The design always begins with a set of given specifications. In this example, suppose the specifications are:

1. Passband ripple of 0 dB ± 0.5 dB from f_1 to f_2, where $f_1 = 3.384$ KHz and $f_2 = 3.664$ kHz.
2. Monotonic stop-band attenuation of at least 39 dB down below 3.100 kHz and above 4.050 kHz.

6.5.2 What if it is not geometric-mean symmetrical? The center frequency f_o of the filter is defined as the square root of the product of the two cutoff frequencies; hence,

$$f_o = \sqrt{f_1 f_2} = 3.52122 \text{ kHz}$$

If the two stop-band frequencies are not geometric-mean symmetrical, i.e., if $\sqrt{f_3 f_4} \neq f_o$, either f_3 or f_4 must be changed to make it symmetrical and still provide more than the attenuation specified. To do this, calculate $f_3' = f_o^2/f_4$ and $f_4' = f_o^2/f_3$. If $f_3' > f_3$, the two symmetrical stop-band frequencies will become f_3' and f_4. If $f_3' < f_3$, the two symmetrical stop-band frequencies will become f_3 and f_4'. In the example, the two −39 dB

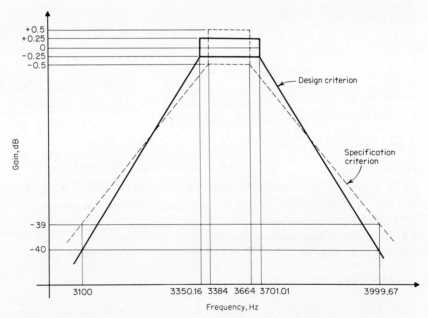

Fig. 6.26 Specification and design criteria of a four pole-pair Tchebysheff bandpass filter.

frequencies become 3.100 kHz and 3.99967 kHz. The dotted plot in Fig. 6.26 illustrates the requirements of the filter.

6.5.3 Allowing guard bands One should not design a filter right to the above specifications. Reasonable guard bands must be allowed to compensate for component tolerances and drifts. How wide the guard bands should be depends on the type of components to be used. The initial component tolerances may be trimmed out on the bench, but the component drifts due to time, temperature, and other environmental conditions should not be ignored. With 741-type amplifiers, 30 ppm/°C NPO ceramic capacitors, and 50 ppm/°C metal film resistors, the passband ripple should be designed to half of that specified, the two cutoff frequencies are each allowed about 1 percent safety margin, and the stop-band attenuation should be designed to be 40 dB instead of the specified 39 dB down. The design limits are plotted in solid lines in Fig. 6.26 and may be stated as:

1. Passband ripple of 0 dB $\pm$ 0.25 dB from 3.35016 kHz to 3.70101 kHz. Note that these two frequencies must be symmetrical about f_o. The first frequency is 1 percent below 3.384 kHz, while the second cutoff frequency is calculated from $f_o^2/3.35016$.
2. Monotonic stop-band attenuation of at least 40 dB down below 3.100 kHz and above 3.99967 kHz.

6.5.4 Estimating number of required poles The next step is to determine the number of pole pairs required. With reference to Sec. 6.3, calculate

$$f = \frac{3.99967 - 3.100}{3.70101 - 3.35016} = 2.56$$

Since the given specifications do not have phase or transient response requirements, Tchebysheff design is used instead of Butterworth to provide the same attenuation rate with fewer pole pairs. Using the nomograph for Tchebysheff filters in Fig. 6.22 and with $A_{min} = 0.5$ dB and $A_{max} = 40$ dB, it is found that four pole pairs are needed.

6.5.5 Normalizing the frequencies In order to find a low-pass prototype and use the transfer-function transformation equations, the center frequency f_o of the filter should be normalized to unity. The normalization implies the division of all frequencies by f_o. Therefore, the two cutoff frequencies become 0.951420 Hz and 1.05106 Hz, while the two −40 dB frequencies become 0.880377 Hz and 1.13588 Hz. The 0.5 dB ($\pm$0.25 dB) bandwidth is 0.09964 Hz.

6.5.6 Finding the low-pass prototype The low-pass prototype, equivalent to the four-pole-pair 0.5 dB ($\pm$0.25 dB) bandpass filter, is the four-pole

0.5 dB Tchebysheff design with cutoff frequency at 0.09964 Hz. The low-pass parameters, as derived from Table 6.3, are

$$\omega_1 = 2\pi\, 0.0594853 \quad Q_1 = 0.705109$$
$$\omega_2 = 2\pi\, 0.102756 \quad Q_2 = 2.94055$$

The low-pass transfer function can, therefore, be written as

$$T(s) = \frac{K_L}{(s^2 + s\omega_1/Q_1 + \omega_1^2)(s^2 + s\omega_2/Q_2 + \omega_2^2)}$$

where K_L is the gain constant, which need not to be determined.

6.5.7 Transformation from low-pass to bandpass

Using the transformation equations given in Sec. 6.4.2, the transfer function of the normalized bandpass filter becomes

$$T(s) = \frac{K_B}{(s^2 + s\omega_a/Q_a + \omega_a^2)(s^2 + s\omega_b/Q_b + \omega_b^2)(s^2 + s\omega_c/Q_c + \omega_c^2)(s^2 + s\omega_d/Q_d + \omega_d^2)}$$

where $\omega_a = 2\pi\, 1.02117$ rad/s $\quad Q_a = 23.7122$
$\omega_b = 2\pi\, 0.979269$ rad/s $\quad Q_b = 23.7122$
$\omega_c = 2\pi\, 1.05189$ rad/s $\quad Q_c = 57.3069$
$\omega_d = 2\pi\, 0.950670$ rad/s $\quad Q_d = 57.3069$
$K_B =$ bandpass gain constant, which need not be determined

The next step is to denormalize the transfer function to its original frequency domain by simply multiplying all frequency terms by f_o. The actual transfer function of the bandpass filter to be synthesized is

$$T(s) = \frac{K_B'}{(s^2 + s\omega_{a1}/Q_a + \omega_{a1}^2)(s^2 + s\omega_{b1}/Q_b + \omega_{b1}^2)(s^2 + s\omega_{c1}/Q_c + \omega_{c1}^2)(s^2 + s\omega_{d1}/Q_d + \omega_{d1}^2)}$$

where $\omega_{a1} = 2\pi\, 3595.76$ rad/s $\quad Q_a = 23.7122$
$\omega_{b1} = 2\pi\, 3448.22$ rad/s $\quad Q_b = 23.7122$
$\omega_{c1} = 2\pi\, 3703.94$ rad/s $\quad Q_c = 57.3069$
$\omega_{d1} = 2\pi\, 3347.52$ rad/s $\quad Q_d = 57.3069$
$K_B' =$ gain constant, to be determined later experimentally

6.5.8 Synthesis using the UAF

Since the Q factors of the filter are relatively high, the universal active filter circuit (UAF) given in Fig. 6.8 should be used because of its low sensitivity. Four UAFs cascaded in series are needed. $C_1 = C_2 = 1000$ pF and $R_3 = K_3 R_3 = 10$ kΩ are the same for all four UAF sections. Sections can be arranged in any order; however, it is advisable to arrange the two high-Q sections in the middle and the two low-Q sections at the first and last sections. This will make the filter more stable with respect to external circuit stresses. With R_4 tentatively equal to 1 kΩ, the calculated element values for the four respective sections are as listed below:

Section	f	Q	$R_1 = R_2$	R_4
1	3,596 Hz	23.7	44.3 kΩ	46.4 kΩ
2	3,704 Hz	57.3	43.0 kΩ	114 kΩ
3	3,348 Hz	57.3	47.5 kΩ	114 kΩ
4	3,448 Hz	23.7	46.2 kΩ	46.4 kΩ

When the four sections are assembled and cascaded together, one will find that the passband gain is excessively high. This is because the center frequency gain of each section is equal to $-(2Q-1)$. It should be noted that the overall filter's passband gain does not equal the product of the four center frequency gains. The passband gain is always much less than that because the four sections are not centered at the same frequency. To drop the gain in any section, simply add a resistor R_5 between ground and the positive summing junction of the input operational amplifier (Fig. 6.8) such that the parallel combination of R_4 and R_5 is equal to 1 kΩ. The gain drop will then be equal to $R_5/(R_4+R_5)$. For high-Q sections, R_4 is normally of the order of 10 to 100 times larger than R_5. Hence when R_4 is adjusted, it changes the gain proportionally but does not affect the center frequency and affects the Q factor only slightly. To keep the filter noises at a minimum, the excessive passband gain should be dropped among the four sections such that when they are cascaded in series, the peak of the gain response at the output of each stage should be around 0 dB. Mathematical derivations can be performed to calculate the values of R_4 and R_5 for each section, but the easy way is to find the values experimentally on the bench because the distribution of gain need not be accurate. Start with the first section, then connect the second section and adjust its gain until the peak of both sections together is around 0 dB, and so on.

6.5.9 Tuning the filter Even with ±1 percent tolerance components, the performance of the filter right after assembly will not be within specifications because the operational amplifiers are assumed to be ideal but actually are not. Operational adjustments must be performed. Unlike passive RLC filters, each active filter section can be adjusted independently. Tune each section as described below:

1. Adjust the center frequency f_o to less than ±1 Hz error. This can be achieved by trimming either R_1 or R_2 and detecting for zero phase shift at f_o.

252 Function Circuits

2. Adjust the Q factor to less than ±2 percent error, where Q is equal to the ratio of f_o to the −3 dB bandwidth.

The last step is to adjust the overall passband gain so as to center around 0 dB. It is not critical in which section the gain is trimmed. Now the filter's response curve should fall within the specification criterion in Fig. 6.26.

6.6 Computer-aided Analysis of Active Filters

To design an active filter with a set of given performance characteristics, the designer needs to first choose an appropriate circuit configuration, then calculate all the component values. The next step is to estimate how stable the filter characteristics would be if one or more of the element values were to change with time, temperature, power supplies, etc. For some simple circuits, sensitivity equations may be derived in terms of characteristic change versus a specific circuit element causing the change. Generally, the mathematical derivations become tedious and unmanageable for even moderately complex circuits. Alternatively, numerical methods may be used to calculate filter sensitivities.

Figure 6.27 lists a computer program named CKTD which will print out the magnitude in decibels and phase in degrees of any active filter network with up to 60 nodes and 100 circuit components, and will also perform the sensitivity analysis numerically. For every circuit component, the program will ask for five input data—A(I1,1) A(I1,2), A(I1,3), A(I1,4) and A(I1.5). I1 is the component number. A(I1,1) and A(I1,2) are the two nodes connecting the component I1, and the current is assumed to flow from A(I1,1) to A(I1,2). A(I1,3) should be assigned 1 if the component I1 is a resistor, 2 if it is a capacitor, 3 if it is an inductor (which it probably never will be for active filters), and 4 if it is a voltage-controlled current source (VCCS). If the network contains other active elements, they must be entered into the program in the form of their equivalent VCCS. A(I1,4) is the actual value of component I1—resistor in ohms, capacitor in farads, inductor in henries, and VCCS in absolute value. A(I1,5) is the control node number if the component is a VCCS and zero if the component is a resistor, capacitor, or inductor. The program will then ask at how many frequencies and at what frequencies the magnitude and phase responses are to be printed out. The number of frequencies cannot exceed 50. In the sensitivity analysis section, the program will ask for a sensitivity factor s. The computer will multiply the nominal component values, one at a time, by S and recalculate the new values for magnitude and phase at a predesignated frequency.

```
1510   370 YB(N1,N1)=YB(N1,N1)+A(I1,4)*B(K5)
1520   GOTO 150
1530   350 YB(N2,N2)=YB(N2,N2)+A(I1,4)*B(K5)
1540   GOTO 150
1550   360 YB(N1,N1)=YB(N1,N1)+A(I1,4)*B(K5)
1560   YB(N2,N2)=YB(N2,N2)+A(I1,4)*B(K5)
1570   YB(N1,N2)=YB(N1,N2)-A(I1,4)*B(K5)
1580   YB(N2,N1)=YB(N2,N1)-A(I1,4)*B(K5)
1590   GOTO 150
1600   180 N1=A(I1,1)
1610   N2=A(I1,2)
1620   IF(N1) 380,390,380
1630   380 IF(N2) 400,410,400
1640   410 YB(N1,N1)=YB(N1,N1)+1./(-B(K5)*A(I1,4))
1650   GOTO 150
1660   390 YB(N2,N2)=YB(N2,N2)+1./(-B(K5)*A(I1,4))
1670   GOTO 150
1680   400 YB(N1,N1)=YB(N1,N1)+1./(-B(K5)*A(I1,4))
1690   YB(N2,N2)=YB(N2,N2)+1./(-B(K5)*A(I1,4))
1700   YB(N1,N2)=YB(N1,N2)-1./(-B(K5)*A(I1,4))
1710   YB(N2,N1)=YB(N2,N1)-1./(-B(K5)*A(I1,4))
1720   GOTO 150
1730   190 IF(A(I1,1)) 210,200,210
1740   210 PRINT 10006
1750   10006 FORMAT (1X, "SOMETHING WRONG IN THE VCCS")
1760   GOTO 10
1770   200 N1=A(I1,2)
1780   N2=A(I1,5)
1790   YA(N1,N2)=YA(N1,N2)-A(I1,4)
1800   150 CONTINUE
1810   MM2=2.*M1
1820   DO 220 K3=1,MM2
1830   DO 230 K4=1,MM2
1840   W(K3,K4)=0.
1850   230 CONTINUE
1860   220 CONTINUE
1870   DO 240 K3=1,MM2,2
1880   DO 245 K4=1,M1
1890   MM3=(K3+1)/2
1900   W(K3,K4)=YA(MM3,K4)
1910   245 CONTINUE
1920   240 CONTINUE
1930   DO 250 K3=2,MM2,2
1940   DO 255 K4=1,M1
1950   MM4=K3/2
1960   W(K3,K4)=YB(MM4,K4)
1970   255 CONTINUE
1980   250 CONTINUE
1990   DO 260 K3=2,MM2,2
2000   MM5=M1+1
2010   DO 265 K4=MM5,MM2
2020   MM6=K4-M1
2030   MM4=K3/2
2040   W(K3,K4)=YA(MM4,MM6)
2050   265 CONTINUE
2060   260 CONTINUE
```

Fig. 6.27 Active filter analysis program CKTD.

```
2070    DO 270 K3=1,MM2,2
2080    DO 275 K4=MM5,MM2
2090    MM3=(K3+1)/2
2100    MM6=K4-M1
2110    W(K3,K4)=-YB(MM3,MM6)
2120    275 CONTINUE
2130    270 CONTINUE
2140    BB(1)=BBB
2150    DO 800 II=2,MM2
2160    800 BB(II)=0.
2170    CALL LINSYS(MM2)
2180    E=DSQRT(BB(MM2)*BB(MM2)+BB(M1)*BB(M1))
2190    E=8.6858896*ALOG(E)
2200    IF(BB(M1)) 605,625,605
2210    625 PH=90.
2220    GOTO 618
2230    605 PH=57.29577*DATAN(BB(MM2)/BB(M1))
2240    618 IF(MN) 68,68,615
2250    615 B(K5)=B(K5)/(2.*3.14159)
2260    PRINT 280,B(K5),E,PH
2270    280 FORMAT (1X,E11.5,4X,E10.4,5X,E10.4)
2280    100 CONTINUE
2290    PRINT 10007
2300    10007 FORMAT(1X,/,1X,"SEN. ANAL., -1 IF NO, FACTOR")
2310    READ/,S
2320    IF(S) 20,25,25
2330    25 PRINT 10008
2340    10008 FORMAT (1X,/,1X,"FREQUENCY")
2350    READ/,B(1)
2360    B(1)=B(1)*2.*3.14159
2370    DO 45 I5=1,M
2380    A(I5,4)=A(I5,4)*S
2390    MN=-1
2400    K5=1
2410    GOTO 55
2420    68 PRINT 75,I5,E,PH
2430    75 FORMAT (1X,I2,2X,E10.4,2X,E10.4)
2440    A(I5,4)=A(I5,4)/S
2450    45 CONTINUE
2460    GOTO 25
2470    END
2480    C- LINEAR SIMULTANEOUS EQTIONS, GAUSSIAN ELIMINATION
2490    SUBROUTINE LINSYS(NAARG)
2500    COMMON W(120,120),BB(120)
2510    DOUBLE PRECISION W,BB,TEMP,TT
2520    1 NA=NAARG
2530    608 DO 291 J1=1,NA
2540    C- FIND REMAINING ROW CONTAINING LARGEST
2550    C- ABSOLUTE VALUE IN PIVOTAL COLUMN
2560    101 TEMP=0.0
2570    DO 121 J2=J1,NA
2580    TT=DAB(W(J2,J1))S
2590    IF(TT-TEMP)121,111,111
2600    111 TEMP=DABS(W(J2,J1))
2610    IBIG=J2
2620    121 CONTINUE
```

Fig. 6.27

```
2630    IF(IBIG-J1)500,201,131
2640  C- REARRANGE ROWS TO PLACE LARGEST ABSOLUTE
2650  C- VALUE IN PIVOT POSITION
2660    131 DO 141 J2=J1,NA
2670    TEMP=W(J1,J2)
2680    W(J1,J2)=W(IBIG,J2)
2690    141 W(IBIG,J2)=TEMP
2700    TEMP=BB(J1)
2710    BB(J1)=BB(IBIG)
2720    161 BB(IBIG)=TEMP
2730  C- COMPUTE COEFFICIENTS IN PIVOTAL ROW
2740    201 TEMP=W(J1,J1)
2750    DO 221 J2=J1,NA
2760    221 W(J1,J2)=W(J1,J2)/TEMP
2770    231 BB(J1)=BB(J1)/TEMP
2780    IF(J1-NA)236,301,500
2790  C- COMPUTE NEW COEFFICIENTS IN REMAINING ROWS
2800    236 N1=J1+1
2810    DO 281 J2=N1,NA
2820    TEMP=W(J2,J1)
2830    DO 241 J3=N1,NA
2840    241 W(J2,J3)=W(J2,J3)-TEMP*W(J1,J3)
2850    251 BB(J2)=BB(J2)-TEMP*BB(J1)
2860    281 CONTINUE
2870    291 CONTINUE
2880  C- OBTAIN SOLUTIONS
2890    301 IF (NA-1)500,500,311
2900    311 N1=NA
2910    321 DO 341 J2=N1,N1
2920    341 BB(N1-1)=BB(N1-1)-BB(J2)*W(N1-1,J2)
2930    N1=N1-1
2940    IF(N1-1)500,500,321
2950    500 RETURN
2960    END
```

6.6.1 An example: eight-pole low-pass Butterworth design with two operational amplifiers   An illustration example, to design and analyze the eight-pole low-pass Butterworth filter using two operational amplifiers and with -3 dB cutoff frequency f_c at 1 Hz, is presented in this section. From Table 6.1, the eight-pole filter parameters for $\omega_c = 1$ rad/s ($f_c = 1/2\pi$ Hz) are, with the lowest Q first, as follows:

$$\omega_1 = 1.00000 \text{ rad/s} \qquad Q_1 = 0.509796$$
$$\omega_2 = 1.00000 \text{ rad/s} \qquad Q_2 = 0.601345$$
$$\omega_3 = 1.00000 \text{ rad/s} \qquad Q_3 = 0.899977$$
$$\omega_4 = 1.00000 \text{ rad/s} \qquad Q_4 = 2.56291$$

The four quadratics ω_1/Q_1, ω_2/Q_2, ω_3/Q_3, and ω_4/Q_4 must be divided into two groups in order to synthesize the eighth-order filter using the four-pole–one-amplifier technique given in Sec. 6.1.9. The first two quadratics ω_1/Q_1 and ω_2/Q_2 are chosen for the first four-pole section, and ω_3/Q_3 and ω_4/Q_4 are chosen for the second four-pole section. The groupings of quadratics are not unique, i.e., other combinations may work as well. Now one needs to find the right capacitors for $f_c = 1$ Hz. This depends on the input bias currents of the two operational amplifiers. If the bias currents are high, the resistors' values should be kept low. That means larger capacitors are needed. $C = 1\ \mu\text{F}$ is used here for both sections.

To calculate the resistor values, recall the computer program WONGA listed in Fig. 6.17a. Run the program twice. The input data for the first run are AF1 = 1.961571 (= $1/Q_1$), F1 = 1., AF2 = 1.66294, F2 = 1., C = 1., R2ST = 1., and R2SP = 1000. The output printout will be R1 = .98426E + 02 (= 98.426 kΩ), R2 = .42112E + 03, R3 = .42112E + 03, R4 = .36759E + 02, and K = .18855E + 01. For the second run, the inputs are AF1 = 1.11114, F1 = 1., AF2 = .390181, F2 = 1., C = 1., R2ST = 1., and R2SP = 1000. The outputs will be R1 = .19595E + 03, R2 = .37914E + 03, R3 = .37914E + 03, R4 = .22779E + 02, and K = .21417E + 01. Note that the overall dc gain of the eighth-order filter would be 4.0382 (the product of 1.8855 and 2.1417). In order to provide unity dc gain, a T network of three resistors, 30.382 kΩ, 90.902 kΩ, and 10 kΩ, is used to replace R1 (= 98.426 kΩ) of the first four-pole section. Figure 6.28a shows the complete circuit diagram of the eight-pole low-pass Butterworth filter with unity dc or passband gain and a -3 dB frequency at 1 Hz.

To prepare for computerized analysis of the filter's performance, the circuit diagram is redrawn in Fig. 6.28b with the following differences from Fig. 6.28a:

1. The nodes are numbered starting from zero, the ground node of the network. Also, each component has a number assigned to it.
2. The input voltage E_i is made equal to 1 V, which is shown in Fig. 6.28b as an input current source of 100 A with source impedance of 0.01 Ω.
3. Each of the two operational amplifiers and its two associated resistors, $(K - 1)\ R$ and R, constitutes a voltage-controlled voltage source (VCVS) with the voltage gain equal to K. Each VCVS is replaced by a voltage-controlled current source (VCCS) with output impedance of 0.01 Ω. The preceding node is the control node.

Figure 6.29 lists the computer printout for both the input and output data. The first and second lines, asking for the number of nodes, the number of components, and the current source input, are self-explanatory. Note that the ground node of zero is not included in counting the number of nodes. The program then asks for "A (1,1), A(1,2),..., A(1,5)........A(M,1), A(M,2),..., A(M,5)?". Five inputs are needed for each circuit component. There are 23 components, and hence a total of 115 numbers is read in. The next line of input data means "(number of frequencies) = 6, F(1) = 0.0001 Hz, F(2) = 0.5 Hz,...., F(6) = 4. Hz." It should be cautioned that the program cannot accept zero hertz. The computer then prints out the gain in

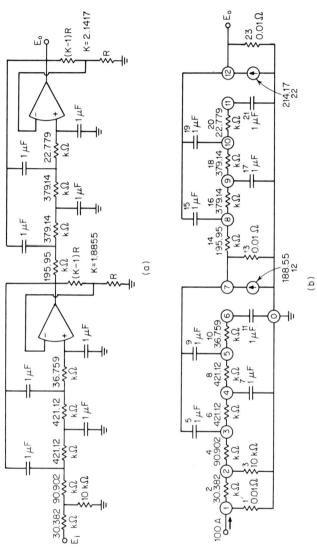

Fig. 6.28 (a) An eight-pole low-pass Butterworth filter with −3 dB frequency at 1 Hz. (b) In order to be analyzed by the program CKTD, VCVS must be converted to its equivalent VCCS.

NO. OF NODES, NO. OF COMPONENTS ?12,23
CURRENT SOURCE INPUT ?100.
A(1,1),A(1,2), . . ., A(1,5)
. . . .
A(M,1),A(M,2), . . ., A(M,5)?1,0,1,1.E-2,0
?1,2,1,3.0382E4,0,2,0,1,1.E4,0,2,3,1,9.0902E4,0
?3,7,2,1.E-6,0,3,4,1,4.2112E5,0,4,0,2,1.E-6,0
?4,5,1,4.2112E5,0,5,7,2,1.E-6,0,5,6,1,3.6759E4,0
?6,0,2,1.E-6,0,0,7,4,1.8855E2,6,7,0,1,1.E-2.0
?7,8,1,1.9595E5,0,8,12,2,1.E-6,0,8,9,1,3.7914E5,0
?9,0,2,1.E-6,0,9,10,1,3.7914E5,0,10,12,2,1.E-6,0
?10,11,1,2.2779E4,0,11,0,2,1.E-6,0,0,12,4,2.1417E2,11
?12,0,1,1.E-2,0

NO. OF FREQ, F(1), . ., F(M2)?6,.0001,.5,.8,1.,2.,4.

FREQ(HZ)	GAIN(DB)	PHASE(DEG)
.10000E-03	-.5517E-04	-.2937E-01
.50000E 00	.3447E-06	.2835E 02
.80000E 00	-.1203E 00	-.8082E 02
.10000E 01	-.3010E 01	.6396E-03
.20000E 01	-.4816E 02	-.2835E 02
.40000E 01	-.9633E 02	.7397E 02

SEN. ANAL., -1 IF NO, FACTOR ?1.01

FREQUENCY?1.

```
 1 -.2923E 01   .6396E-03
 2 -.3076E 01  -.5517E-02
 3 -.2947E 01  -.1816E-01
 4 -.3050E 01  -.3007E 00
 5 -.3041E 01  -.3804E 00
 6 -.3085E 01   .1463E-01
 7 -.3073E 01   .6965E-01
 8 -.3031E 01  -.4489E 00
 9 -.2735E 01  -.7519E 00
10 -.3044E 01  -.5064E 00
11 -.3354E 01  -.2217E 00
12 -.2305E 01  -.7815E 00
13 -.2305E 01  -.7815E 00
14 -.3252E 01  -.1334E 01
15 -.3268E 01  -.2149E 01
16 -.3289E 01   .1771E 01
17 -.3272E 01   .3788E 01
18 -.2540E 01  -.1394E 01
19  .1200E 01  -.2016E 01
20 -.3132E 01  -.3061E 01
21 -.5609E 01  -.3274E 01
22  .7973E 01  -.2378E 01
23  .7973E 01  -.2378E 01
```

FREQUENCY?

Fig. 6.29 Input and output data of analyzing the eight-pole low-pass Butterworth filter.

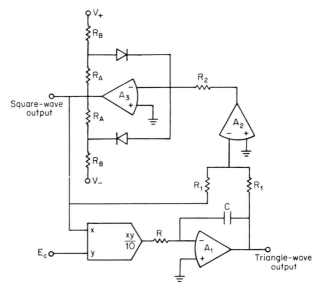

Fig. 7.1 This voltage-controlled waveform generator may also be used as an analog-to-pulse-rate converter, a sweep generator, a voltage-controlled oscillator, or a voltage-to-frequency converter.

tional amplifiers and one multiplier. Amplifier A_1 serves as an integrator, while amplifiers A_2 and A_3 form a comparator circuit. The positive feedback around the comparator provides fast, uniform switching when the oscillation frequency is changing. The simple diode limiter (R_A, R_B, and the two diodes) around amplifier A_3 sets the amplitude of the square-wave output. With the multiplier connected in the feedback loop, the control voltage E_c tunes the amplitude of the square wave feeding back to the integrator. The multiplier acts as a frequency modulator in the circuit, which would otherwise oscillate at a frequency determined by the RC product of the integrator. The frequency of the output square wave and triangle wave is given by

$$f_o = \frac{E_c}{40RC} \qquad E_c > 0$$

By cascading one of the two-quadrant sine-function generators described in Chapter 5 on the triangle-wave output, an ultra-low-frequency sine-wave output can be obtained.

7.1.2 Voltage-controlled quadrature oscillator[1] A quadrature oscillator consists of two integrators and some sort of amplitude-limiting circuit. With the addition of two low-cost multipliers, the frequency of oscillation can be made linearly proportional to a control voltage. The principle of operation may be derived from the linear-voltage-tunable active filter given in Sec. 6.1.6. Rewrite the bandpass transfer function of the filter circuit in Fig. 6.11,

$$\frac{E_{BP}(s)}{E_i(s)} = \frac{K_{BP}s\omega_0/Q}{s^2 + s\omega_0/Q + \omega_0^2}$$

where $\omega_0 = \dfrac{E_c}{10RC}$

$Q = \dfrac{R_2 + R_3}{2R_2}$

$K_{BP} = \dfrac{R_3}{R_2}$

When R_3 is made infinite, Q becomes infinite and the poles of the transfer function will move to the $j\omega$ axis. The circuit will, therefore, oscillate with input grounded. Mathematically, when $R_3 = \infty$, the transfer function simplifies to

$$\frac{E_{BP}(s)}{E_i(s)} = \frac{2s\omega_0}{s^2 + \omega_0^2}$$

With input grounded, i.e., $E_i(s) = 1$, the inverse Laplace transform of $E_{BP}(s)$ is

$$e_{BP}(t) = 2\omega_0 \cos \omega_0 t$$

Similarly, it can be derived that the oscillation at the low-pass output terminal has the form

$$e_{LP}(t) = 2\omega_0 \sin \omega_0 t$$

In practice, some kind of amplitude-limiting circuit is needed to prevent the oscillation from diverging into saturation. Automatic gain control techniques may also be used. Figure 7.2a depicts the complete circuit diagram of a voltage-controlled quadrature oscillator with a simple diode limiter. The oscillator circuit is the same as the active filter circuit in Fig. 6.11 except that:

1. R_3 in the filter circuit is infinite.
2. R_2 in the filter circuit is not used here; it may be included but will have no practical effect.
3. A limiter circuit, made up of two diodes and four resistors, is added.

The amplitude of oscillation is determined by the ratio of R_A to R_B in the limiter. With ± 15 V power supplies and ± 10 V oscillator output swing,

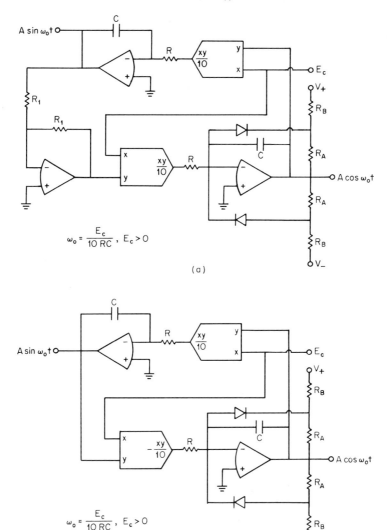

Fig. 7.2 (a) and (b) Complete voltage-controlled quadrature oscillators. The circuit in (b) requires one inverting multiplier.

typical values of R_A and R_B may be 10 kΩ and 15 kΩ, respectively. R_A should always be made much smaller than R.

The inverting amplifier in Fig. 7.2a can be omitted if an inverting multiplier is available. The circuit connections are illustrated in Fig. 7.2b. Any differential input multiplier may be connected as an inverting multiplier.

7.1.3 Voltage-variable time constant

A multiplier may be used to remotely and dynamically control a time constant without affecting the gain. The desired transfer function is

$$\frac{E_o(s)}{E_i(s)} = \frac{K}{\tau s + 1} \quad \text{where } \tau \text{ is a variable}$$

The circuit in Fig. 7.3 can be used to vary the time constant over a wide range independent of the gain K. The transfer function in terms of circuit elements is

$$\frac{E_o(s)}{E_i(s)} = \frac{-R_2/R_1}{(10R_2C/E_c)s + 1}$$

From the above two equations, it becomes obvious that

$$K = -\frac{R_2}{R_1} \quad \text{and} \quad \tau = \frac{10R_2C}{E_c}$$

For most commercial multipliers operating from ± 15 V power supplies, $(E_c)_{max} = 10$ V; then $\tau_{min} = R_2C$. As E_c decreases, τ increases. The circuit accuracy is excellent for large E_c, but for very small E_c the multiplier error becomes significant. A 10:1 range of variation in τ is readily achievable

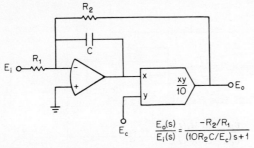

Fig. 7.3 Voltage-variable time constant circuit. This circuit may also be used as a simple-pole voltage-controlled low-pass filter.

with general-purpose multipliers. This circuit may also be used as a simple-pole low-pass active filter whose cutoff frequency is linearly proportional to the control voltage E_c.

7.1.4 Voltage-controlled exponentiator and antilog amplifier

The multifunction converter discussed in Sec. 3.6 performs the computation of

$$e_o = v_y \left(\frac{v_z}{v_x}\right)^m$$

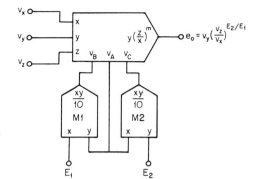

Fig. 7.4 Versatile voltage-controlled exponentiator.

where v_x, v_y, and v_z are input voltages, and m is an arbitrary exponent determined by two external resistors. By employing two multipliers instead of two resistors, the exponent m can be made voltage-tunable. As illustrated in Fig. 7.4, the transfer function is

$$e_o = v_y \left(\frac{v_z}{v_x}\right)^{E_2/E_1} \tag{7-1}$$

where E_1 and E_2 are control voltages.

Equation (7-1) may be derived by referring back to the basic theory of the multifunction converter, discussed in Sec. 3.6.1. Rewrite Eqs. (3-49) and (3-50):

$$v_R = \frac{kT}{q} \ln \frac{i_z}{i_x} \tag{7-2}$$

$$v_C = \frac{kT}{q} \ln \frac{i_o}{i_y} \tag{7-3}$$

If Eq. (7-3) is divided by (7-2), the resulting equation is

$$\frac{v_B}{v_C} = \frac{\ln (i_z/i_x)}{\ln (i_o/i_y)} \tag{7-4}$$

From Fig. 7.4, $v_B = v_A E_1/10$ and $v_C = v_A E_2/10$. The ratio of v_B to v_C

$$\frac{v_B}{v_C} = \frac{E_1}{E_2} \tag{7-5}$$

Eliminate v_B/v_C from Eqs. (7-4) and (7-5):

$$\frac{E_1}{E_2} = \frac{\ln (i_z/i_x)}{\ln (i_o/i_y)}$$

or

$$i_o = i_y \left(\frac{i_z}{i_x}\right)^{E_2/E_1} \tag{7-6}$$

266 Function Circuits

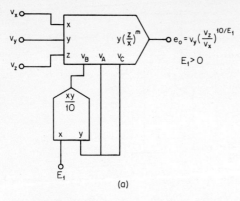

(a)

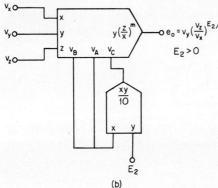

(b)

Fig. 7.5 Voltage-controlled exponentiator using one multifunction converter and one analog multiplier. (a) $e_o = v_y (v_z/v_x)^{10/E_1}$, where $E_1 > 0$. (b) $e_o = v_y (v_z/v_x)^{E_2/10}$, where $E_2 > 0$.

From Fig. 3.26, $i_o = e_o/R_o$, $i_y = v_y/R_y$, $i_z = v_z/R_z$, and $i_x = v_x/R_x$. If $R_o = R_y = R_z = R_x$, Eq. (7-6) may be converted into the voltage transfer function:

$$e_o = v_y \left(\frac{v_z}{v_x}\right)^{E_2/E_1} \qquad (7\text{-}7)$$

Two simplified special cases may be derived from Eq. (7-7). In the first case, where $E_2 = +10$ V, the multiplier $M2$ forces $v_c = v_A$. This is the same as shorting v_A and v_C together, and the multiplier $M2$ can be saved. Figure 7.5a depicts the circuit diagram using one multifunction converter and one multiplier. The transfer function is

$$e_o = v_y \left(\frac{v_z}{v_x}\right)^{10/E_1}$$

Similarly, the second case, where $E_1 = +10$ V, imposes $v_B = v_A$. The multiplier $M1$ can be saved by simply shorting v_A and v_B together. The circuit diagram in Fig. 7.5b provides the transfer function of

$$e_o = v_y \left(\frac{v_z}{v_x}\right)^{E_2/10}$$

Equation (7-7) may also be considered as an antilog operation. Let us make $v = v_z/v_x$ and $E = E_2/E_1$ just to simplify the analysis. Now we have

$$e_o = v_y v^E$$

or

$$\frac{e_o}{v_y} = v^E$$

Taking the logarithm of both sides to the base v, the equation becomes

$$\log_v \frac{e_o}{v_y} = E$$

This results in the antilog operation of

$$e_o = v_y \text{ antilog}_v E$$

7.2 Modulators and Frequency Doublers[4,5]

Modulation and frequency doubling are basically processes of multiplication. Technically, modulation is described as a process by which some characteristics of one wave, called a carrier, are varied in accordance with some characteristics of another wave, called a modulating signal. In the past, modulators were not generally designed using multipliers, but nowadays, low-cost integrated circuits have made it practical to design modulators with multipliers.

7.2.1 Balanced modulator
Balanced modulators are widely used in telecommunication equipment, measurement instruments, and control systems. They have the advantage over other modulation schemes that the carrier is suppressed and does not appear in the output. Thus power consumption is reduced.

As illustrated in Fig. 7.6, an analog multiplier, with two ac signals applied to its two inputs, is actually performing balanced modulation, or suppressed-carrier double-sideband modulation. Analytically, if the two inputs are sinusoidal, the output is

$$e_o = \frac{1}{10}(E_m \sin \omega_m t)(E_c \sin \omega_c t)$$

where 1/10 is the multiplier gain constant, ω_m is the modulating signal frequency, and ω_c is the carrier frequency. This equation can be expanded as

$$e_o = \frac{E_m E_c}{20} [\cos(\omega_c - \omega_m)t - \cos(\omega_c + \omega_m)t]$$

The carrier-frequency term does not appear in the above equation, and thus the name *suppressed-carrier* is obtained.

Most other modulation circuits have their spectrum centered about the second harmonic of the carrier frequency, or any multiple of it, and re-

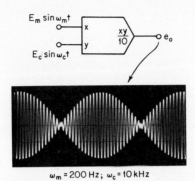

Fig. 7.6 Use of an analog multiplier as a balanced modulator.

quire complex filters to eliminate the unwanted frequencies. With two sinusoidal input signals, the multiplier contains only two frequencies in its output, the sum and the difference of the two input frequencies. The filter requirements are minimal.

A common problem in communications is to extract information from single-sideband (ss) signals received. The ss signal can be written in the form

$$e_{ss} = K \sin(\omega_m + \omega_c)t$$

If e_{ss} is multiplied by an appropriate carrier signal, $A \sin \omega_c t$, the resulting output will be

$$v_o = \frac{1}{10} [K \sin(\omega_m + \omega_c)t](A \sin \omega_c t)$$

$$= \frac{KA}{20} [\cos \omega_m t - \cos(\omega_m + 2\omega_c)t]$$

The intelligence contained in the first term, $(KA/20) \cos \omega_m t$, can be extracted easily by using a simple filter to remove the second high-frequency term.

7.2.2 Amplitude modulator When a dc voltage is added to the modulating signal, the multiplier performs amplitude modulation in a way similar to the balanced modulation. This procedure allows the carrier to pass through the multiplier when the modulating signal is zero. As shown in Fig. 7.7, the output of the multiplier is

$$e_o = \frac{1}{10}[E_m + mE_m \sin \omega_m t](E_c \sin \omega_c t)$$

$$= \frac{E_m E_c}{10}\left[\sin \omega_c t + \frac{m}{2}\cos(\omega_c - \omega_m)t - \frac{m}{2}\cos(\omega_c + \omega_m)t\right]$$

where m = modulation index

It is easy to achieve 100 percent modulation by simply making the peak amplitude of the modulating wave equal to the dc offset voltage. With most commercial multipliers, input feedthroughs (sometimes called input offsets) are adjustable. Adjusting the feedthrough error voltage of the in-

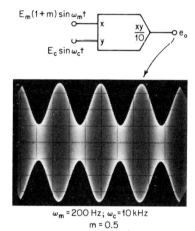

Fig. 7.7 If a dc voltage is added to the modulating signal, an analog multiplier may operate as an amplitude modulator.

put where the ac modulating signal is applied is equivalent to externally summing the ac modulating signal with a dc voltage.

7.2.3 Frequency doubler

A diode is traditionally used to double the frequency of an ac signal. However, the fundamental frequency plus a series of harmonics are also generated with this method, and extensive filtering is required to extract the desired harmonic. Also, the amplitude of the second harmonic obtained this way is usually small and needs additional amplification.

If the two input terminals are connected together, as shown in Fig. 7.8,

270 Function Circuits

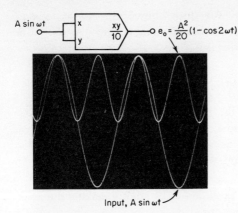

Fig. 7.8 By connecting the two inputs together, an analog multiplier may operate as a frequency doubler.

the multiplier operates as a frequency doubler. The principle is based on the equations

$$e_o = \frac{1}{10}(A \sin \omega t)^2$$
$$= \frac{A^2}{20}(1 - \cos 2\omega t)$$

The output of the multiplier contains a dc voltage in association with the second harmonic of the input signal. The dc voltage can be removed through ac coupling.

7.3 Phase-Angle Detector

The phase difference between two signals of the same frequency can be detected using an analog multiplier in series with a low-pass filter, as

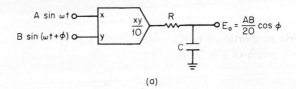

(a)

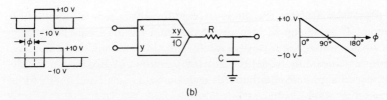

(b)

Fig. 7.9 A simple phase detector, containing one multiplier in series with a low-pass RC filter, to measure the phase difference between two sine waves in (a) and between two square waves in (b).

illustrated in Fig. 7.9. In the first case (Fig. 7.9a), where both signals are sinusoidal waves of $A \sin \omega t$ and $B \sin (\omega t + \Phi)$, respectively, the principle of phase detection is based on the trigonometric identity

$$\sin \alpha \sin \beta = \frac{1}{2} [\cos (\alpha - \beta) - \cos (\alpha + \beta)]$$

The waveform at the output of the multiplier consists of an ac voltage superimposed on a dc level. The frequency of the ac term is twice that of the input signals, while the dc voltage level is proportional to the phase difference between the two inputs. Mathematically,

$$\text{Multiplier output} = \frac{xy}{10}$$

$$= \frac{AB}{20} [\cos \Phi - \cos (2\omega t + \Phi)]$$

The low-pass RC filter eliminates the ac term while retaining the dc voltage output E_o, where

$$E_o = \frac{AB}{20} \cos \Phi$$

or

$$\Phi = \text{arc} - \cos \frac{20 E_o}{AB}$$

In the second case (Fig. 7.9b), where both inputs are square waves, the resulting voltage waveform at the output of the multiplier will be a pulse train whose duty cycle is in direct proportion to the phase-angle difference Φ of the two input square waves. The duty cycle is 100 percent when $\Phi = 0°$, 50 percent when $\Phi = 90°$, and 0 percent when $\Phi = 180°$. The output low-pass filter retains only the dc component of the pulse train which is linearly proportional to the phase angle Φ, as illustrated in the voltage-versus-Φ graph of Fig. 7.9b.

7.4 The Phase-locked Loop[2,3]

The phase-locked loop (PLL) is a very useful feedback circuit with a variety of applications in instrumentation and communication systems. The original concept was developed in the 1930s, but it did not become practical until recently, with the availability of low-cost integrated circuits. As shown in Fig. 7.10, the basic PLL is made up of a phase detector, a low-pass filter, and a voltage-controlled oscillator (VCO).

The phase detector is basically an analog multiplier and works on the principle discussed in Sec. 7.3. The theory of operation of the PLL can

272 Function Circuits

be understood by first considering the input terminal grounded; then the error voltage $\epsilon(t)$ at the output of the phase detector and therefore at the output of the low-pass filter is zero. The VCO operates at a preset, free-running frequency f_o determined by its internal RC product. When an input signal $e_i(t)$ is applied to the PLL, the phase detector compares the phase and frequency of the input signal with the VCO and generates an error voltage $\epsilon(t)$. The low-pass filter extracts the dc content of $\epsilon(t)$, which is applied to the VCO. If the frequency of $e_i(t)$ is in the vicinity of f_o, the feedback loop will force the VCO to synchronize, or lock, on the input signal. Once locked, the frequency of the VCO is the same as that of the input signal except for a finite phase difference.

Applications of the PLL may be grouped in two fundamentally different classes. In the first class, the PLL is used as a demodulator, following phase or frequency modulation. The PLL may be considered as a matched filter operating as a coherent detector. In the second class, the PLL is used to track a carrier or synchronizing signal which may vary in frequency with time. The PLL, in this class, may be thought of as a narrowband filter with a high degree of frequency selectivity to remove noise from a signal.

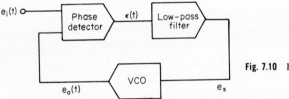

Fig. 7.10 Basic phase-locked loop.

7.5 Polar-to-Rectangular Resolver[1]

Resolvers are often used in fire control systems, navigation computers, and general-purpose analog computers to convert polar coordinates to rectangular coordinates. For example, some types of radar provide targeting information in the form of a vertical angle and a distance to the target, which must then be converted to rectangular coordinates. Polar-to-rectangular resolvers require the computation of

$$x = R \cos \theta$$
$$y = R \sin \theta$$

Various methods of generating sine and cosine functions can be found in Chapter 5. Figure 7.11 shows the interconnections of sine generators, cosine generators, and analog multipliers to compute rectangular coordinates (x,y) which can then be displayed on a CRT screen. Before

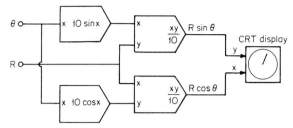

Fig. 7.11 Generation of CRT displays.

choosing a sine or cosine generator, be sure that it will operate in the quadrants required by your specific application, since there are one-quadrant and two-quadrant as well as four-quadrant sine/cosine generators. One-quadrant sine/cosine generators will operate only with inputs from 0 to +90°, two-quadrant generators will operate from −90° to +90°, and four-quadrant generators will operate from −180° to +180°.

7.6 Automatic Gain Control Circuits[1,6]

Automatic gain control (AGC) circuits are widely used to stabilize the signal amplitude of oscillators, to keep a signal's amplitude constant while its phase angle is varied by filtering, and for various other purposes. It is natural to think of the analog multiplier for automatic gain control because it can be considered a voltage-controlled linear amplifier. The block diagram of an AGC circuit is shown in Fig. 7.12. The amplitude of the output signal e_o is first rectified and then filtered to retain its dc component. The dc voltage at the output of the low-pass filter is compared with a reference voltage through the integrator to generate an error signal. The error signal is integrated in the high-gain integrator. When the dc component of the output voltage e_o is equal to the reference voltage, the integrator input is zero and the output of the integrator is steady. The output of the integrator is multiplied by the input signal e_i, thus varying the gain. The amplitude stability of the output signal depends primarily

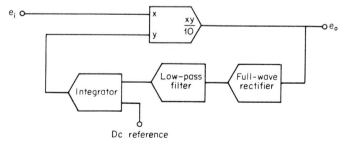

Fig. 7.12 Block diagram of an automatic gain control circuit.

on the stability of the dc reference voltage and on the integrator gain. The rate at which the loop can restabilize after sudden changes in input level depends on the cutoff frequency of the low-pass filter and the integrator time constant.

7.6.1 Simple AGC using two-quadrant dividers Automatic gain control circuits need not be limited to employing multipliers as control elements. Simple AGC circuits using two-quadrant dividers are shown in Fig. 7.13; the numerator of the dividers will accept bipolar voltages, but the denominator in (a) accepts only positive voltages and the denominator in (b) accepts only negative voltages. The feedback circuits work in a manner very similar to the circuit using multipliers shown in Fig. 7.12. The output voltage e_o is half-wave rectified through the diode and then compared to a dc reference by an integrator. The integrator's output voltage is applied to the denominator of the divider as the control signal. Therefore, variations in the output e_o caused by input voltage changes will create a control signal that compensates for the input voltage changes.

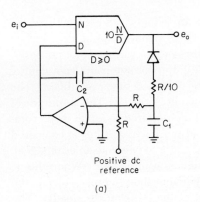

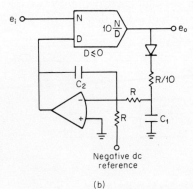

Fig. 7.13 Simple automatic gain control circuits employing two-quadrant dividers. The divider in (a) accepts only positive denominator voltages, and the one in (b) only negative denominator voltages.

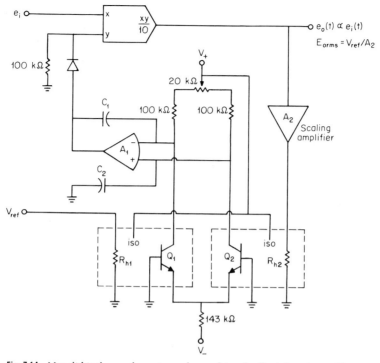

Fig. 7.14 Monolithic thermoelements can be used in a feedback loop to provide very wideband automatic gain control, or rms voltage regulation.

7.6.2 Wideband automatic gain control[7]

The monolithic thermoelements described in Chapter 4, Sec. 4.3, can be used in place of the full-wave rectifier in Fig. 7.12 to provide very wideband AGC. The AGC circuit shown in Fig. 7.14 uses the monolithic thermoelements, and in addition to wideband, precision automatic gain control, it regulates the rms signal level independent of the input waveform or harmonic distortion. In this circuit, the output signal's rms value is compared to a dc reference with a pair of monolithic thermoelements, and a feedback signal is generated which is used to control the gain of a wideband multiplier. Capacitors C_1 and C_2 provide loop-phase compensation and set the low-frequency cutoff for regulation. The value of these capacitors is a function of the scaling amplifier's gain (or attenuation), frequency response, and the low-frequency cutoff desired.

7.7 Gain Measurement

Measurement of an amplifier's gain, gain nonlinearity, and frequency response are some of the most common measurements made. The device under test need not necessarily be an amplifier as such; it may be a multiplier, divider, log amplifier, or some other function circuit, but under some specific conditions it is often desired to test its transfer frequency response, or its "gain" linearity. Presented in this section are two gain-measuring circuits; one has the particular advantage of rms response with very wideband capability, while the other is particularly suited to measuring an "amplifier's" low-frequency gain and gain nonlinearity.

7.7.1 Wideband gain measurement[7]   A stage's gain can be determined by measuring the input voltage and then the output voltage; the gain, of course, is simply the ratio of the output voltage to the input voltage. If a high degree of precision is required in an ac gain measurement, rms-responding measurements should be made. The rms-responding ac measurement is necessary since phase shift of any harmonics in the input signal due to finite bandwidth in the stage being tested can cause significant errors in an average- or peak-responding measurement. For instance, only 1 percent third-harmonic distortion in a signal can produce as much as 0.33 percent error, depending on the phase shift, in an average-responding measurement. An rms-responding ac-to-dc converter, on the other hand, responds to the total signal independent of the relative phase of its components.

A precise gain-measuring circuit can be built around a pair of monolithic thermoelements which are both rms-responding and very wideband. A suitable circuit is shown in Fig. 7.15, and a thorough discussion of the monolithic thermoelements is given in Chapter 4, Sec. 4.3. In this circuit the input to the amplifier, or stage, under test is also applied to one of the monolithic thermoelements, and the output of the amplifier is applied to the other thermoelement through an appropriate attenuator. Any difference between the rms values of these signals causes differential heating of the two thermoelements. The resulting differential voltage is then amplified by the differential stage, consisting of the temperature-sensing transistors Q_1 and Q_2 and their load resistors and the operational amplifier A_1. This amplified error signal is applied through an attenuator to the base of the inverting transistor to provide negative feedback. The output of the operational amplifier is then an amplified replica of the error signal. The capacitor C_1 reduces the gain on the ac error signal, providing a dc output for even very low frequency inputs.

Analysis of the feedback loop yields the dc output voltage

$$E_o = \frac{R_2 + R_1}{R_1} A_{T01} \left[\frac{A_{T02}}{A_{T01}} \left(\frac{V_{\text{OUTrms}}}{K'} \right)^2 - E_{\text{irms}}^2 - \frac{V_{BE2} - V_{BE1}}{A_{T01}} \right]$$

where A_{T01} and A_{T02} are the dc thermal gains of the input and output thermoelements, respectively, and $V_{BE2} - V_{BE1}$ is the ambient mismatch

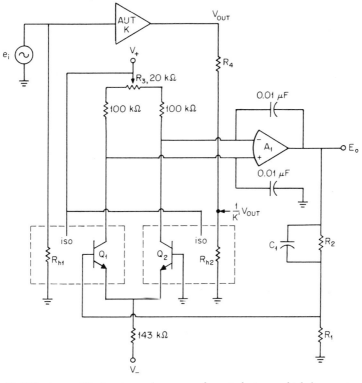

Fig. 7.15 An amplifier's gain can be measured precisely to very high frequencies using this thermal-responding, null-seeking feedback loop.

between the base-emitter voltages of the two temperature-sensing transistors. (For a complete discussion of these parameters of the thermoelements, see Chapter 4, Sec. 4.3.) There are two primary sources of measurement error in the output, mismatch in the ambient base-emitter voltages and mismatch in the thermal gains of the thermoelements. The difference in base-emitter voltages can be nulled by adjusting R_3 for zero output voltage with zero input signal. The mismatch in the thermal gains can be compensated by adjusting the attenuation factor K for

$$K' = \frac{R_{h2}}{R_{h2} + R_4} = \sqrt{\frac{A_{T01}}{A_{T02}}}\, K$$

by applying an ideal signal at V_{OUT} and adjusting R_4 for zero output.

If the error in the amplifier under test's gain δ is small, the calibrated output simplifies to

$$E_o = \frac{2A_{T01}(R_1 + R_2)}{R_1} \delta E_{\text{irms}} \qquad \text{where } \delta = \left(\frac{V_{\text{OUT}}}{K'} - E_{\text{irms}}\right) \ll 1$$

278 Function Circuits

For a specific signal level, and small gain error in the amplifier or stage being tested, the output of the gain-measuring circuit is essentially a linear function of the stage gain error.

The upper frequency limit on this test circuit is primarily that of the attenuator at the output of the amplifier under test. But also, capacitive coupling of high-frequency swing into the dc circuit, particularly at the inputs of the operational amplifier A_1, can produce errors in the output level. These high-frequency measurement error sources are discussed in Chapter 4, Sec. 4.3.5.

7.7.2 Low-frequency gain error and nonlinearity measurement One must often measure an amplifier's gain accuracy at low frequencies, and its gain nonlinearity may be of particular interest. A circuit which automatically nulls the amplifier under test's gain error and provides amplification of the gain nonlinearity is shown in Fig. 7.16. This circuit requires the amplifier being tested to have a negative gain for operation; if the stage has a positive gain, a precision inverter can be inserted at its output. The input to the amplifier is derived on a precision attenuator equal to the inverse of the ideal gain. The output of the amplifier can then be compared to the input of the attenuator. Since the amplifier is inverting, this can be easily accomplished with a matched pair of resistors R_3 and R_4

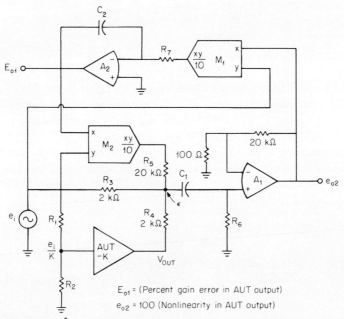

Fig. 7.16 The low-frequency gain and nonlinearity error of an amplifier or stage can be precisely measured with this null-seeking feedback loop.

connected between the attenuator input and the amplifier output. The voltage at the junction of these two resistors would be zero if the amplifier's gain perfectly matched the attenuation of the input. Any deviation in the output from the ideal appears attenuated by a factor of 2 at this junction, and is the voltage ϵ in Fig. 7.16.

The error voltage ϵ is amplified by a gain of 200 V/V to produce the output e_{o2}. As will be seen, e_{o2} is the stage being tested's gain nonlinearity in a gain of 100 V/V. This voltage is correlated with the input voltage by M_1 and the integrator composed of A_2, R_7, and C_2 to produce a dc level which is only related to the component of ϵ which has the fundamental frequency. By multiplying this dc level with the input, a gain correction signal can be supplied to the junction of R_3 and R_4, thereby nulling the fundamental component of signal present in ϵ. The error signal ϵ then contains only the harmonics of the input frequency and is, therefore, only due to the nonlinearity in the AUT output voltage. The output e_{o2} is then an amplified replica of the gain nonlinearity, and the output E_{o1} is a dc level which is linearly related to the AUT's gain error.

Another way of analyzing this circuit is to recognize that E_{o1} can be made a dc level by choosing a sufficiently long integration time constant R_7C_2. By multiplying this dc level with the input signal and applying it to the junction of the resistors R_3 and R_4, a change in the amplitude of the fundamental component of ϵ can be made, but only in the fundamental. Multiplying the error voltage ϵ with the input ensures that the input to the integrator has a dc component related to the fundamental component in ϵ. Maintaining the proper phase relationship provides negative feedback on the fundamental component of ϵ, thereby nulling the fundamental component in ϵ. This results in E_{o1} being linearly related to the AUT's gain error,

$$E_{o1} = \frac{10R_5}{R_4}\left(1 + \frac{V_{OUT}}{e_i}\right)$$

$$= 100\left(1 + \frac{V_{OUT}}{e_i}\right)$$

For example, a +1.0 percent gain error results in $E_{o1} = +1$ V dc. The output e_{o2} is an ac voltage which is the AUT's output nonlinearity amplified 100 times, so a 1 mV p-p nonlinearity in the AUT output results in $e_{o2} = 100$ mV p-p.

In order to eliminate any effects of dc offset voltage in the amplifier, the voltage ϵ is ac-coupled to the times 200 stage through C_1. The time constant R_6C_1 must be long enough to pass the second harmonic of the input without significant error. Figure 1.23 in Chapter 1 is useful for choosing this time constant. The multiplier M_1 has only a minimal effect on the measurement accuracy, since in the steady state the fundamental is nulled to zero, and there will be very little direct current at its output. The integration time constant should ensure that the ripple in the output E_{o1} is small. This ripple is due to the nonlinearity and should be long enough to filter the second harmonic. Gain errors and gain nonlinearity in multiplier M_2 can cause significant measurement errors. The errors in the output are reduced by the ratio R_5/R_4, and this should be kept as large as possible. A final note: phase shift of the fundamental in the AUT will produce large errors in the nonlinearity measured at e_{o2}, and the frequency of the input should be low enough to produce negligible phase shift. Again, the graph in Fig. 1.23, Chapter 1, is useful. In this case the vector error curve is pertinent; if, for example, a nonlinearity of 0.05 percent is to be measured, the vector error should be much less than this.

7.8 Power Measurement

Generally there are two classes of power measurements: measuring the power being delivered to the load, and measuring the power capable of being delivered to a specific load. An example of the first class would be monitoring the power being consumed by a system in use, and an example of the second would be testing the output power capability of a signal generator with a specific test load impedance. The first parts of this section discuss single- and three-phase power-measuring circuits suitable for continuous monitoring of the power consumed. The last part considers a circuit suitable for testing the output power capability of a signal source with a specific load.

7.8.1 Single-phase power measurement Single-phase power being delivered to a load is easily measured with a multiplier and a low-pass filter. If the product of the load voltage and current is found, the instantaneous power is obtained:

$$p(t) = e_i(t) i_L(t)$$

For sinusoidal signals,

$$e_i(t) = \sqrt{2} E_{irms} \sin \omega t$$

$$i_L(t) = \sqrt{2} I_{Lrms} \sin (\omega t + \theta)$$

the instantaneous power is

$$p(t) = 2 E_{irms} I_{Lrms} \sin \omega t \sin (\omega t + \theta)$$

Use of the identity

$$\sin x \sin y = \frac{1}{2} \cos (x - y) - \frac{1}{2} \cos (x + y)$$

allows the power to be rewritten as

$$p(t) = E_{irms}I_{Lrms}[\cos\theta - \cos(2\omega t + \theta)]$$

Note that the instantaneous power has two components, a dc component equal to the real power and an ac component with a frequency twice that of the input fundamental. Therefore, if the instantaneous power is low-pass filtered to extract the dc component, the real power is obtained.

$$\text{Real power} = E_{irms}I_{Lrms}\cos\theta$$

The power factor is $\cos\theta$, and can be written as

$$\cos\theta = \frac{\text{real power}}{E_{irms}I_{Lrms}}$$

The power factor can be computed with this expression by dividing the real power determined above by the apparent power, which is the product of the rms values of the input voltage and load current. These can be developed using the rms converters discussed in Chapter 4.

7.8.2 Three-phase power measurement[8] The three-phase power being delivered to a load can be measured with three circuits, one measuring the real power in each leg (the circuit in Figure 7.17 would be suitable);

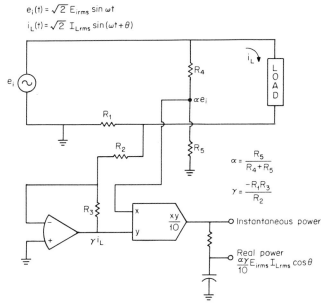

Fig. 7.17 A multiplier and a low-pass filter compute both the instantaneous power and the real power being delivered to a load.

simply sum these three levels to obtain the total power delivered. However, the low-pass filtering used to extract the dc component in these circuits will limit the speed of response. Examination of the sum of the instantaneous powers, P_T, reveals that the low-pass filters are unnecessary with three-phase measurements and balanced loads:

$$P_T = 2E_{irms}I_{Lrms}\left[\sin \omega t \sin(\omega t + \theta) + \sin\left(\omega t + \frac{2\pi}{3}\right)\sin\left(\omega t + \frac{2\pi}{3} + \theta\right)\right.$$
$$\left. + \sin\left(\omega t + \frac{4\pi}{3}\right)\sin\left(\omega t + \frac{4\pi}{3} + \theta\right)\right]$$

$$= 3E_{irms}I_{Lrms}\cos\theta$$

Note that no harmonics are present in the sum of the instantaneous powers and no filtering is required. The circuit implementation of the three-phase power measurement is shown in Fig. 7.18.

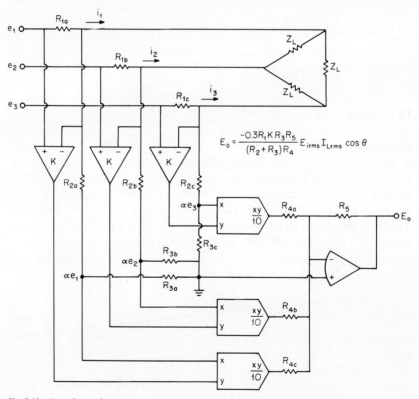

Fig. 7.18 Fast three-phase power measurement can be made without filtering.

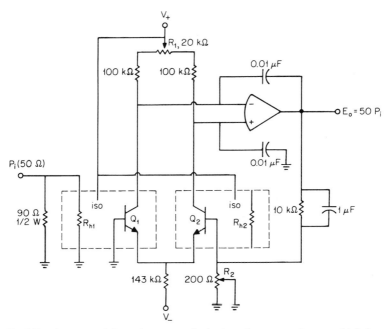

Fig. 7.19 The power delivered to a specific load can be measured to very high frequencies with this thermal rms-responding measurement circuit.

7.8.3 Wideband power measurement[7] The power delivered to a specific load can be measured to very high frequencies using a pair of monolithic thermoelements (the monolithic thermoelement was discussed in Chapter 4, Sec. 4.3). The circuit in Fig. 7.19 converts the power delivered to a specific load to a dc output voltage and is particularly useful for testing an amplifier's or signal generator's rms power output capability. This circuit is scaled to 50 Ω input impedance. By adjusting R_1 for zero output with zero input, and R_2 for an output of 9 V dc with 3 V dc input, accuracies within ±2 percent to 100 MHz are possible. The high-frequency limitations are the same as those discussed for the thermal rms converter in Chapter 4, Sec. 4.3.5.

7.9 Mass Gas Flow Computation

Designers in the process industry can save time and money by using general-purpose function circuits as building blocks to solve complex control problems. One example is to use two low-cost multifunction converters to obtain direct measurements of gas flow through an orifice. The equation of mass gas flow is given by

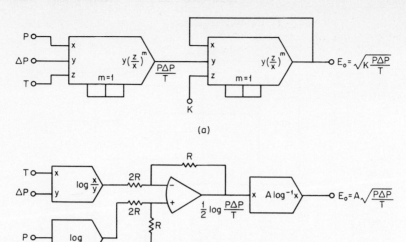

Fig. 7.20 Compute mass gas flow (a) using two multifunction converters and (b) using logarithmic and antilog amplifiers.

$$E_o(\text{flow}) = \sqrt{K \frac{P \, \Delta P}{T}}$$

where ΔP is the differential pressure across the orifice, P is the absolute pressure, T is the absolute temperature, and K is a constant scaled to utilize the full output range. The equation can be implemented as shown in Fig. 7.20a.

If the variables (ΔP, P, and T) vary over a wide dynamic range, the circuit in Fig. 7.20b, employing logarithmic amplifiers, is more appropriate because it compresses the input signal levels and eliminates the problem of over-ranging as long as the output voltage is properly scaled by the gain constant A of the antilog amplifier. A logarithmic ratio amplifier is utilized to take the log ratio of $T/\Delta P$.

REFERENCES

1. T. Cate and H. Handler, Designing with Packaged Analog Multipliers, *EEE*, May 1969.
2. A. B. Grebene, The Monolithic Phase Locked Loop: A Versatile Building Block, *EDN*, Oct. 1, 1972.
3. T. B. Mills, The Phase Locked Loop IC as a Communication System Building Block, *AN-46* National Semiconductor Corporation, Santa Clara, Calif.
4. *Specifications and Applications Information—MC 1595L, MC 1495L*, Motorola Semiconductors, Phoenix, Ariz.
5. *Specifications and Applications Information—MC 1594L, MC 1494L*, Motorola Semiconductors, Phoenix, Ariz.

6. L. Counts and D. Sheingold, Caution: Care Needed When Using Lowest Cost Analog Dividers, *EDN*, April 20, 1974.
7. W. E. Ott, Monolithic Converter Augments AC-Measurement Capabilities, *Electronics*, Jan. 23, 1975.
8. L. R. Smith and L. A. Snider, Detection and Measurement of Three-Phase Power, Reactive Power, and Power Factor, with Minimum Time Delay, *Proc. IEEE*, November 1970.

INDEX

Ac-to-dc conversion, 126
  average-responding, 128
  peak-responding, 127
  [See also Rms (root-mean-square) conversion]
Active filters, 213–259
  active twin T, 221
  all-pass, 226
  band-reject, 220
  bandpass, 217
    center frequency tuning, 219
  computerized analysis of, 252
  converting bandpass to band-reject, 220
  fourth order with one op amp, 232
  geometric-mean symmetrical, 217, 220, 244
  high-pass, 216
  low-pass, 95, 214
  multiple feedback, 215, 219
  multistage design example of, 247
  notch (see band-reject above)
  state-variable, 219, 250

Active filters (Cont.)
  synthesis of, 214
  third order with one op amp, 228
  tuning, 251
  universal (see state variable above)
  voltage-controlled, 225, 264
  voltage-controlled voltage source, 214, 216
Amplifiers:
  antilog, 61, 264
  composite, 164, 167
  difference (see Difference amplifiers)
  emitter-degenerated, 96
  inverting (see Inverting amplifiers)
  log-ratio, 63
  logarithmic (see Logarithmic amplifiers)
  noninverting, 5, 13, 167
  operational (see Operational amplifiers)
  ranging, 164
  summing, 3, 6, 14, 34
  voltage-controlled linear, 273

Index

Amplitude, 126
 average value, 127, 128
 peak value, 127
 rms value, 127, 129
Amplitude modulators, 268
Analog dividers (*see* Dividers, analog)
Analog multipliers (*see* Multipliers, analog)
Antilog amplifiers, 61, 264
Approximation with noninteger exponent, 201
Arbitrary-function generators, 190
Arc-tangent circuit, 204
Automatic gain control, 273
 using dividers, 274
 wideband, 275

Balanced modulators, 267
Bessel filters, 236, 237
Bias current (*see* Input bias current)
Bode plot, 23, 54, 76
Boltzmann's constant, 97
Broadbanding, 25, 166
Buffering, input, 60
Bulk resistance error, 48
Butterworth filters, 236, 237
 nomograph for, 243

Capacitive loads, 33
Carrier, 267
Coherent detectors, 272
Common-mode rejection, 18, 43
 measurement of, 19
Common-mode voltage, 18
Composite amplifiers, 164, 167
Converters (*see* Multifunction converters)
Cosine generators, one quadrant, 203
Crest factor, 127, 169
Current inverters, 63
Current source, voltage-controlled, 252, 256

Delay at cutoff, 239
Demodulators, 272
Detectors:
 coherent, 272
 peak, 127
 phase-angle, 270
Difference amplifiers, 5, 20, 34, 42, 43
Differential amplifiers (*see* Difference amplifiers; Instrumentation amplifiers)
Differentiators, 39

Diode-function generators, 209
Diode limiters, 90
Dividers, analog, 107–113
 adjustment procedures of, 112
 converting one quadrant to two quadrant, 110, 113
 feedthrough of, 112
 linearity compensation of, 112
 log-antilog, 110
 multiplier-inverted, 109, 119
 offsets of, 112
 one quadrant, 104, 108, 110, 117
 two quadrant, 108
Double-sideband modulation, 267
Drift, 11
 of input bias current, 11
 of offset voltage, 11

Ebers-Moll equation, 46, 147
Equal ripple (Tchebyscheff approximations), 182, 237
Exponentiators, 119, 264

Feedback factor, 12
Filters:
 active (*see* Active filters)
 brick-wall, 237
 ideal low-pass, 237
 nomographs for, 243
 passive *RLC*, 213
Form factor, 127
Frequency compensation (*see* Phase compensation)
Frequency doublers, 267, 269
Frequency modulation, 272
Frequency response:
 of differentiators, 40
 of integrators, 39
 of log amplifiers, 55, 68
 of operational amplifiers, 23
 of rms converters, 137, 142, 153, 160, 162, 172
Full-power bandwidth, operational amplifier, 29
Function circuits, applications of, 260
Function generators, 177

Gain error, 13, 275
Gain measurement, 275
 low-frequency, 278
 wide-bandwidth, 276
Gain nonlinearity measurement, 278
Gas-flow computation, mass, 283
Gaussian elimination method, 191

Index 289

Generators, 177
  arbitrary-function, 190
  cosine, one quadrant, 203
  diode-function, 209
  inverse function, 200
  sine (*see* Sine generators)
  square-wave, 260
  triangular-wave, 95, 260
  waveform, 177
    voltage-controlled, 260

Implicit feedback, 207
Infinite series, 177
Input bias current, 7, 59
  compensation, 60
  drift versus temperature, 11, 60
  FET buffer, 60
  measurement, 12
  noise, 11
Input difference current, 9, 11
Input impedance:
  of instrumentation amplifiers, 43
  of operational amplifiers, 21
Input offset voltage, 7
  adjustment, 9, 61
  drift versus temperature, 11
  measurement, 12
  noise, 11
Input resistance (*see* Input impedance)
Instrumentation amplifiers, 42
  ideal response, 42
  input impedance of, 43
  model, 42
Integrated circuit (IC) multiplier, 82, 85, 96
Integrators, 37
  effects of input offsets, 38
  frequency response of, 39
Inverse function generators, 200
Inverters, current, 63
Inverting amplifiers, 2, 13, 164, 166

Kawakami, M., 243

Laplace transform, 262
Linear phase (Bessel) filters, 237
Log-antilog divider, 110
Log-antilog multiplier, 86, 102
  adjustment procedures for, 104
  linearity compensation of, 104
Log-ratio amplifier, 63
Logarithmic amplifiers, 46, 52
  bipolar transistor, 46
  log-conformity compensation, 48, 52

Logarithmic amplifiers (*Cont.*)
  log-transistor protection, 55
  phase compensation, 52
  selecting operational amplifier, 59
  specifications of, 65
  temperature effects of, 57, 68
  testing, 68
Logarithmic multiplier (*see* Log-antilog multiplier)
Loop gain, 13
Low-pass prototype, 245, 249

Magnitude error, 28
Maximally flat (Butterworth) filters, 237
Modulating signal, 267
Modulation index, 269
Modulators, 267
  amplitude, 268
  pulse-height, 90
  pulse-width, 90
Multifunction converters, 71, 113, 121
  applications of, 201
  practical, 116
  theory of operation of, 114
Multipliers, analog, 71, 107
  accuracy of, 74, 82, 87
  adjustment procedures for, 84, 104
  applications of, 260
  Bode plot of, 76
  comparison of different types of, 84
  converting one quadrant to four quadrant, 104, 106
  crossplotting (*see* testing *below*)
  differential phase shift of, 82
  distortion of, 76, 79
  dynamic performance of, 76
  error of (*see* accuracy of *above*)
  feedthrough of, 74, 79
  frequency response of, 76, 87
  ideal, 72
  linearity of, 73, 79
  log-antilog (*see* Log-antilog multiplier)
  nonlinearity of (*see* linearity of *above*)
  null suppression of, 74, 79
  offsets of, 73
  one quadrant, 102, 117
  phase shift of, 76, 77
  practical, 72
  settling time of, 77
  slew rate of, 77
  specifications of, 72
  testing, 82
  vector error of, 77

Index

Natural frequencies, 213
Nomographs, filter, 243
Noninverting amplifiers, 5, 13, 167
Null techniques (*see* Input offset voltage, adjustment; Input bias current, compensation)

Offset current (*see* Input difference current)
Offset voltage (*see* Input offset voltage)
Open-loop gain, 12, 166
   measurement of, 14
Operational amplifiers:
   composite, 166
   dc limitations of, 7
   dynamic characteristics of, 23
   ideal properties of, 2
   summing point restraints, 3
   symbol of, 2
Oscillator, quadrature, 261
Output impedance:
   of instrumentation amplifiers, 43
   of operational amplifiers, 16
Output resistance (*see* Output impedance)

Passive *RLC* filters, 213
Peak detectors, 127
Phase-angle detector, 270
Phase compensation, 24, 52
   for broadband, 25
   for capacitive loads, 33
   of log amplifiers, 52
   selection of, 25
Phase constant, 239
Phase-locked loop, 271
Phase margin, 24, 54
Polynomial generation, computer-aided, 192
Power factor, 281
Power measurement, 280
   single-phase, 280
   three-phase, 281
   wide-bandwidth, 283
Power series, 278
   realization of, 278
Protection circuits:
   for log amplifiers, 55
   for thermal rms converters, 160
Pulse-height modulator, 91
Pulse-width modulator, 91
Pulse-width/pulse-height multiplier, 85, 90
   externally excited, 91
   internally excited, 91

Quadrature oscillator, voltage-controlled, 261
Quarter-square multiplier, 84, 87

Ranging amplifier, 164
Real power, 281
Resolver, polar to rectangular, 272
Rms (root-mean-square) conversion:
   computing rms converter, 131, 141
   crest factor, 127, 169
   high-frequency limitations, 143, 160
   low-frequency errors, 137, 153, 162
   settling time, 140, 158, 164
   step response, 139, 156
   thermal rms converter, 144
      protection circuits for, 160
   (*See also* Ac-to-dc conversion)
Rms (root-mean-square) level regulation, 275

Sensitivity analysis, 252, 259
Series, 177
Settling time:
   of log amplifiers, 55, 69
   of operational amplifiers, 31
   of rms converters, 140, 158, 164
Sine generators:
   four quadrant, 207
   one quadrant, 201
   two quadrant, 179, 190
Single-sideband signals, 268
Slew rate:
   effects on pulse train, 174
   measurement, 29
   of operational amplifiers, 29
Square rooters, 122–125
   divider-converted, 123
   multiplier-inverted, 122
   using multifunction converter as, 124
Square-wave generator, 260
Squaring circuits, 85, 87
State-variable technique, 219, 250
Summing amplifiers, 4, 6, 14, 34
Summing junction, 3
Superposition, law of, 8
Suppressed-carrier modulation, 267
Switch, electronic, 93

T feedback, 4
Taylor's series, 177
   truncated, 186
Tchebyscheff filters, 237, 240
   nomograph for, 243

Tchebyscheff polynomials, 179
  coefficients of, 181
  computerized expansion into, 186
  expansion into a series of, 182
  truncated, 186
Thermal gain, 148
Thermal time constant, 148
Thermocouples, type E chromel-constantan, linearization of, 196
Thermoelements, 145, 275, 276, 283
  monolithic, resistor-transistor, 146
  resistor-thermocouple, 144
Time-constant circuit, 264
Time delay, 238
Time-division multiplier (see Pulse-width/pulse-height multiplier)
Transconductance multiplier, 82, 85, 96
  linearity compensation of, 102
Transfer functions, transformation of, 244
Transformation, 244
  low-pass to band-reject, 247
  low-pass to bandpass, 246, 250
  low-pass to high-pass, 245

Transient protection (see Protection circuits)
Triangle-averaging multiplier, 85, 93
Triangular-wave generator, 95, 260
Twin T networks, 221

Unity-gain bandwidth, 25

Variable-transconductance multiplier (see Transconductance multiplier)
Vector error, 28
Vector-magnitude circuit, 204
Voltage-controlled current source, 252, 256
Voltage-controlled voltage source, 256

Waveform generators, 177
  voltage-controlled, 260